K

THE PILLBUG PROJECT

A guide to investigation

This book has been edited and produced by the staff of NSTA Special Publications.

Library of Congress Catalog Card Number 92–60247

Stock Number PB–93

ISBN Number 0-87355-109-5

Printed in the United States of America

THE PILLBUG PROJECT

A guide to investigation

by Robin Burnett

illustrated by Sergey Ivanov

National Science Teachers Association

The Author

My name is Robin Burnett and I was born in 1945. My father, a professor of science education, taught me early that dirty hands lead to clean science. A B.S. from the University of Illinois and a Ph.D. in biology from Stanford added a hint of respectability. A decade of teaching experimental biology at Stanford helped me realize that too few get the needed early training in investigative science. To help motivate people to explore their world, my wife and I joined forces with two others and dreamed up the Monterey Bay Aquarium. Next we started a company, Sea Studios, which makes natural history videos. Right now I am building a barn and being a husband to one and father to three.

The Illustrator

I was born in June 1957 and was named Sergey. Moscow is my hometown, so I am Russian. My last name is a comon one over there: Ivanov. It's like "Johnson" in this country. I studied at the Moscow Physical-Technological Institute, but then I discovered I would rather paint pictures than solve equations. I studied with Oscar Kacharov, a professor at the Moscow Institute of Fine Arts, who helped me learn the mysteries of watercolor. Now am an artist with the National Science Teachers Association. I live in Arlington, Virginia, with my wife (a violinist) and son (a kindergartener).

Acknowledgments

MANY PEOPLE ASSISTED in the development of this book. Chief among them were our reviewers, who provided terrific suggestions and advice. Bonnie B. Barr is a professor of science education at State University of New York at Cortland, New York. Dr. Barr is 1991–92 president of the Council for Elementary Science International. She has taught high school biology, been an elementary science specialist, is senior author for a junior high textbook series, and has written numerous articles in professional journals.

David C. Kramer, is a professor of biology and science education at St. Cloud State University, St. Cloud, Minnesota. Dr. Kramer has been *Science and Children* column editor for "The Classroom Animal," and has published several articles on using pillbugs in the classroom.

Karen K. Lind is a professor of early childhood and middle childhood education at University of Louisville, Kentucky. Dr. Lind was 1990–91 president of the Council for Elementary Science International. She is *Science and Children* column editor for the "Early Childhood" column.

Arthur Livermore has substantial experience in science education. As a professor of chemistry at Reed College in Oregon he directed the NSF-funded high school curriculum project known as Chemical Bond Approach. In the 1970s he helped direct the NSF-funded elementary curriculum project Science—A Process Approach.

Alice Moses is currently a program director in the education and human resources division of the National Science Foundation. Dr. Moses is a past president of the National Science Teachers Association.

The Pillbug Project was produced by NSTA Publications, Shirley Watt Ireton, managing editor; Christine Marie Pearce, assistant editor; Andrew Saindon, assistant editor, Gregg Sekscienski, production editor; and Anne Marie Calmes, editorial assistant. Christine Pearce was the NSTA editor for *The Pillbug Project*. The book was designed by Thomas Mann of Mann & Mann Graphic Design and Jane D'Alelio of Icehouse Graphics.

Table of Contents

Preface

AS IS TYPICAL with most projects, I suppose, the introduction is being written last. Indeed, in this case it has been four years since the Pillbug Project was first put together and tested. I originally wrote *The Pillbug Project* for my son Jason and his second grade classmates. Simply put—they loved it. Months later the children were still coming up to me to describe some "neat thing" they found out about pillbugs.

Since then the Pillbug Project has been modified and tested several times, most recently at the Beach School in Lummi Island, Washington. The Beach School has 50 students total in grades 1 through 6, and the entire school participated in the project. The first graders were given handouts written to fit their reading abilities, but other than that, the project was presented to all the students together. Once again—they loved it. The project seems to have enough flexibility that students of a wide range of abilities and interests can all be challenged by it, gain from it, and enjoy it.

No, the students didn't all turn into scientists after the project was over, or even actively study pillbugs. But they did learn to look at pillbugs, and they learned a bit about different ways to look at things around them and about experimenting in general. More importantly, though, they had a chance to discover and investigate on their own. They experienced what science is really about—learning directly from the world around us.

Introduction

YOU'LL FIND THAT the Pillbug Project works with a minimum of equipment and preparation time. While these features may ease your task each day of getting ready, their main purpose is to let the students get involved directly with nature.

Probably the most important feature of the day-to-day workings of the Pillbug Project is that it walks the middle ground between the possible boredom of orderly "cookbook science" found in many classrooms and the probable chaos that would result if you told students to go out and be scientists but didn't lead them through the process. The Pillbug Project is intended to introduce the *ways* of science, not give established information or teach specific procedures.

Science is, after all, an exploratory process. Science's flexible structure has components that are surefire tools to capture the interest of elementary students: coming up with a reasonable question to ask of nature, observing and manipulating nature to find the answer to the question, formulating an explanation for the results of those observations and manipulations, and then finally, attempting to prove wrong that explanation.

Every step of this process follows logically from the previous steps. Though there may be several options for each step, scientists—including student scientists—develop a logical reason for the steps they select for their procedure. If your students can begin asking their own questions about nature and begin learning to follow logical steps instead of just written "cookbook" instructions to direct their activities—in this project and in subsequent activities—then the Pillbug Project will have been successful.

The Pieces of the Pillbug Project

There are three sections in each day of the Pillbug Project: Teacher Notes, Teacher Narrative, and Research Notes. Teacher Notes will give you background on the activity and suggestions for organizing student activity and setting up the day's materials. Possible discussion questions are included to help you guide students' curiosity and thinking. These notes should help everything run smoothly.

Teacher Narrative will guide your presentation to students. It is written in such a way that you could just read it aloud like any other written material—but if you do this, step aside occasionally and emphasize that certain phrases are silly or serious or make a good point or whatever. The purpose of the Teacher Narrative is to give you a general idea of what needs to be said and a sense of the classroom atmosphere I intend for the project. As author, I have my own personality and way of speaking, my own interests and biases. Add or subtract as you will to what I have provided in this book to tailor the Pillbug Project to your students. You know your class and have your way of running things and should decide where to set up, what specific pieces of equipment to use, who to team up with whom, how long to run a particular portion, and so forth.

The pages of Research Notes are intended to be reproduced as handouts. Young students may not yet recognize the logic of the steps of their activities without guidance, so Research Notes includes written instructions for your students to guide them through the Pillbug Project's activities, particularly during the first few days. Beyond this minimal "cookbook" function, the pages of Research Notes also serve as a logbook. These pages reinforce the concepts of experimental design, give students a record of their investigation, serve as a take-off point for further exploration by students, and are a form of portfolio assessment that you and the students can use to gauge how much they have learned. Encourage students to come up with their own questions for their log books, and to add additional pages if they need or want them.

There is enough room along the left margins of the Research Notes for the students to bind the pages together to create an actual notebook. The drawings they make for the *Patricia Pillbug* story would work nicely as covers for their research notebooks. The more the project has the stamp of official business, the more the children will get out of it. And the more easily paperwork can be organized, filed, and retrieved, the more likely the children will use and gain from the experience.

A central element of the Pillbug Project is a story that you read aloud to the class, the *Patricia Pillbug* story, about the adventures of a pillbug—named Patricia, of course. It is in three installments for reading during three days of the project, and it provides the impetus for two activities. It is a fun story with some humorous parts and some perilous parts and some illustrations that illuminate Patricia's world. It is an invitation to consider the world from a different perspective. But it also has a touch of a moral: pillbugs are living creatures and they deserve respect as such. This point should be emphasized throughout the Pillbug Project and generalized to include any living creature they might use in any science project. Living creatures should be handled with great care, minimizing chances of unnecessary harm. But this is not meant to put a damper on the story. It should be read for fun and will, hopefully, spark children's imagination and enthusiasm.

Conducting and Evaluating the Pillbug Project

Students will become involved with their individual investigations or projects and will want things to work. As teacher, your task will be to facilitate whatever they are trying to do and to give them advice when they need it. While it is they who are doing the work, you need to see that they aren't doomed to frustration. Here you must be the tightrope walker—giving them the freedom to try what they want, but helping them to avoid trying things that, in your judgment, are not reasonable.

Encourage your students to work in teams throughout the Pillbug Project. This will cut down on the number of pillbugs and the amount of equipment you need, and the cooperative learning will simulate the work environment of real research laboratories where scientists collaborate on projects. But for successful cooperative learning, teams need to be kept small enough to make sure that no student can sit back and let the others do everything. At the start, you may want to tell students that their group has two tasks—to explore pillbugs and to make sure each member of the group participates.

Several methods for assessing the success of the Project and the knowledge of your students are embedded in the Project. The log book is a form of portfolio assessment that you and the students can use. In looking over the log books, students will be able to see relationships among the activities, and see the quality of their work and their understanding improve.

On Sixth Day and Tenth Day, students give group reports of what they have discovered. In the Pillbug Project, these cooperative reports are called seminars, and serve to let students educate each other in the same way scientists do, by explaining their methods and sharing their results.

On Seventh Day, students share their reactions to what life would be like as a pillbug with questions that relate a pillbug's form to its habits and habitat, and which explore beginning concepts in adaptation. Their method is called a gallery walk, another form of cooperative assessment.

Pillbug Fact Sheet

Like adult scientists, your young scientists are bound to discover that working to answer one question often spawns several more questions. To help answer some of those questions, I have provided a Pillbug Fact Sheet for your reference. Some of the information the students will discover on their own, but other bits of information are beyond the scope of this project. You might copy this sheet after the project is over for students to add to their research folders and encourage them to add facts that the class discovered over the course of the project. With some luck, your scientists will come up with questions that aren't answered by the fact sheet or by the project or possibly by anyone. Then they will have experienced the driving force of science!

Habitat facts

- Pillbugs are not insects but are crustaceans, like lobsters, crabs, and crayfish. They are in a subgroup called *isopods*.
- 4,000 species of isopods have been identified so far.
- Most isopods live in marine habitats, some live in fresh water, and very few types live on land.
- Of those isopods that live on land, some can roll up into a ball, or *pill* shape, hence the term *pill*bug. But the term is used generically to refer to all types of land isopods.
- Other common names are: sow bug, potato bug, wood lice, and roly-poly.
- *Hikers* and *rollers* are terms used in this book to distinguish two types of pillbugs that respond differently when disturbed. Hikers start moving faster, whereas rollers roll up into a ball.
- Pillbugs breathe through gill-like structures and therefore must live in moist places.
- Pillbugs are usually found in dimly lit or dark places.
- Pillbugs live underneath things that provide darkness and moisture (such as under rocks, boards, leaf litter, tree bark, etc.).
- In their natural habitat, pillbugs generally eat decaying wood, leaves, and other vegetation.
- The young are self-sufficient when born.

More Pillbug Facts

Anatomy and Physiology Facts

- Pillbugs have three body regions: head, thorax, and abdomen.
- A pillbug is 5–15 mm long.
- They have seven pairs of legs, all of which are identical.
- Most of the body is covered with shield-like plates.
- A pillbug's "skeleton" is a hard outer skin (the shield-like plates are part of this skin).
- Pillbugs are cold blooded. Their body temperature depends on the temperature of their surroundings.
- Females may produce up to 200 eggs which are carried under the thorax in a sack-like thing called a brood pouch. They will reproduce more often during warmer temperatures.
- The young look like miniature adults.
- They may roll into a ball when disturbed or to lessen water loss when humidity is low.
- Pillbugs molt, shedding their old skin and producing a new one. How often they molt depends on temperature, diet, and other factors.

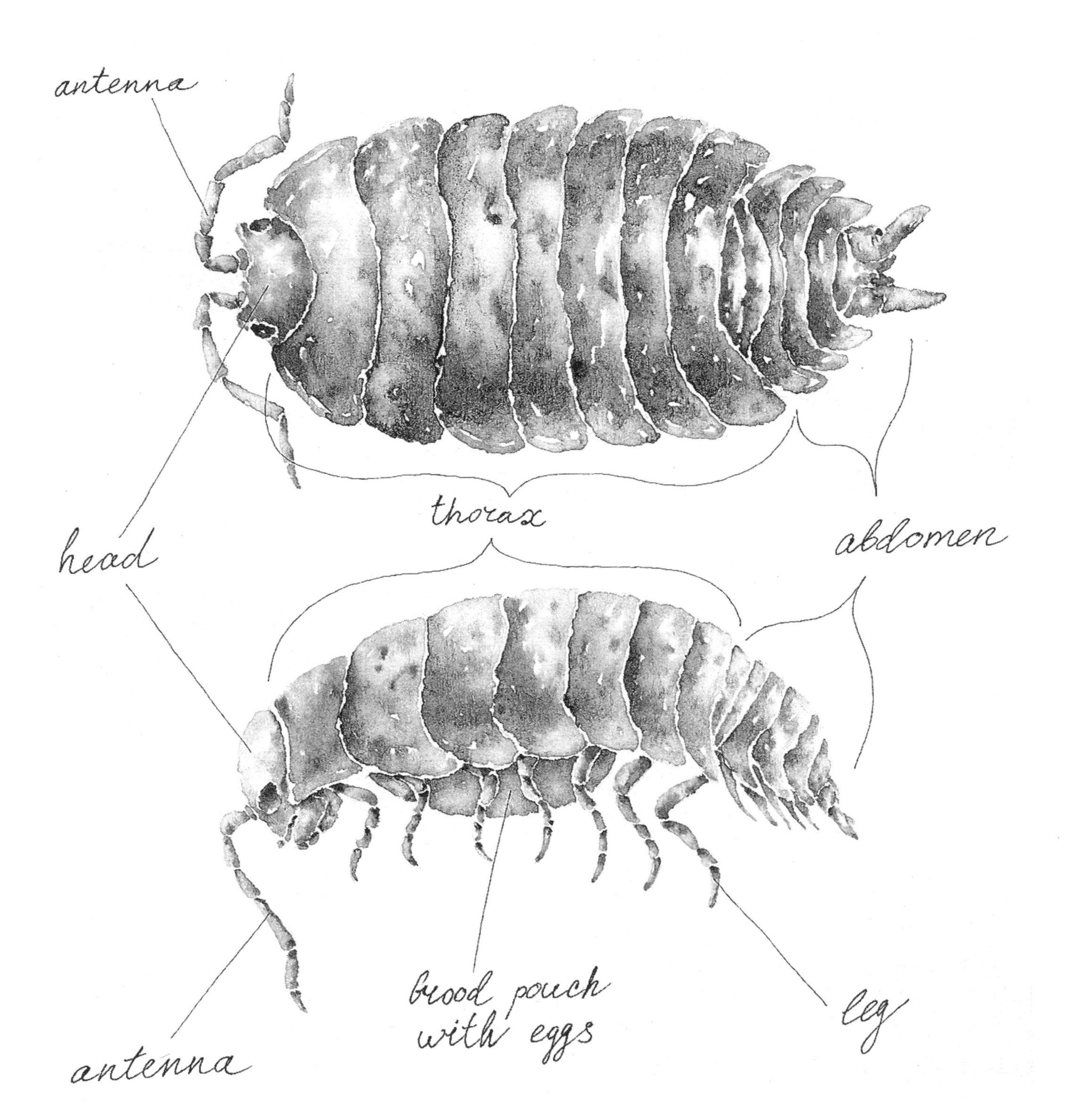
antenna
head
thorax
abdomen
antenna
brood pouch with eggs
leg

Preparing for the Pillbug Project

Perhaps the most important thing in preparing for the Pillbug Project is to make sure that there are enough pillbugs near your school—enough so that each student can find five or six during a few-minute hunt. If you aren't sure that there are ample pillbugs hiking around near your school, turn over a few rocks, look under logs and other objects, or look carefully in flower gardens. If you can find a dozen or so pillbugs after 10 minutes of looking, there probably are enough for the project. But if your school yard is particularly dry and has very few places where pillbugs can hide, you had better plan on modifying the Project. Perhaps you could take the children on a field trip for the outside activities and import pillbugs from elsewhere for your classroom investigations. Pillbugs can be purchased from a number of biological supply houses.

If you do use the school yard as an outdoor laboratory, you may want to advise the grounds staff or custodian about your activities. They might even be able to give you some hints as to where pillbugs could be found.

The other major requirement in preparing for the Pillbug Project is to make sure there is enough time to run it. The way it is currently structured, it takes about 90 minutes each day for ten days. You might be able to use a slightly shorter time period, but if a time crunch seems likely in your classroom, consider making the same span of time fill several needs. You could integrate other outdoor and ecological studies with the project or emphasize the language arts aspects of the project.

Certainly, the whole project could run longer. Results from several of the experiments could be used to spawn a second generation of experiments that follow up the first set of experiments. This second set of experiments could be interesting, since the students would be using what they found in the first set to make their second set tell them even more.

First Day

Teacher Note

As you can see from the materials list, you need little for this first day. The children will be going outside to find out where pillbugs live, so all they need is the appropriate outdoor clothes. At the end of the search, you will ask them to bring back some pillbugs for the pillbug hideaway.

Materials List

- A pillbug hideaway—The hideaway can be any container which has a lid or a top (you will need to put a few airholes in the top if it doesn't already have them). Plastic sweater boxes, plastic tubs, wash basins with window screen lids—anything like this will work fine. The pillbugs can stay here between investigations.
- Some dirt, pine needles, grass, or mulch—You should put a handful or so (enough to generously cover the bottom) into the hideaway.

- A dash of water—You need this to moisten—not soak—whatever you have put into the bottom of the hideaway. The pillbugs need moisture to breathe.
- A small handful of dry oatmeal, bread crumbs, or dry cereal—Pillbugs eat decaying vegetation in their natural habitat, but these substitutes will be palatable enough for them.
- Plastic spoons—Pillbugs are fairly robust, but students should be taught to handle them carefully. Have students pick them up gently with a spoon.
- Paper cups—Your scientists can put collected pillbugs into these.

Teacher Note

This is a fun day. The students are going to look for where pillbugs live. It will seem a lot like an Easter egg hunt, with children running around looking for pillbugs. Invariably they will get excited at a particularly good find, and they will share it with everyone.

Show students how to lift stones and sticks carefully, and to roll them back into place after they've looked beneath them. Remind students that even if a pillbug doesn't live under an object, other creatures might, and they should minimize any disturbance to creatures' living quarters.

Some children won't know what a pillbug is, or at least won't know it by that name. Show them some pillbugs you have found or a picture from a book. Tell children enough for them to recognize pillbugs when they find them, but leave the physical details for them to discover on their own. You'll want to describe a pillbug for them and use whatever local names you might know. On Lummi Island in Washington the children all know them by potato bug—which is a name I use for a cricket-like insect. Near Monterey, California they call them roly-polies. Common names usually come from appearance or behavior. You could ask your students "If you were to invent a name for the pillbug, what would it be?"

As for the initial pillbug hunt, it is probably better to assign the children to small groups of three or four, so they begin to combine their efforts and they don't scatter to the winds. But when it's time to collect pillbugs for the hideaway, each student should collect his or her own—just a few depending on how common they are. Before they begin be sure to tell them not to remove all of the pillbugs from any one place. They don't want to decimate the local population. Leave a few behind to carry on.

This hunt has two parts. First, students will discover how common pillbugs are, then they will begin to look at the characteristics of the areas where pillbugs are most often found.

The Research Note page will add structure to their search. If you read the questions aloud to them (or have older students read the questions) ahead of time and encourage them to work toward answering the questions, you will help them focus on the pillbugs. But once they are out actively hunting them, it may be a bit intrusive to insist that they actually write out a detailed answer to each question. You might want them to record their observations and answer the questions on the first page of their research notebook after they return from their hunt.

I have found that getting the children together for a group recap of their hunt is fun. After they have finished hunting, call them in. They will all be enthusiastic about their particular discoveries. To build their enthusiasm further, point out to them that since they are seeing particular pillbugs that no one else has ever studied or probably even seen, they might find out something that no one else knows. Take a couple of minutes to let them talk about anything strange about the pillbugs they saw: different kinds, different sizes, some with babies and such. Encourage generalizations as well. You may want to ask "How would you describe the shape of a pillbug?" "Which pillbugs do you think are related to each other?" "How are all of the pillbugs similar?" "If you were a pillbug, where would you hide to get away from the sun and still get moisture?" Draw a picture of your hiding place. "What might you want to eat if you were a pillbug?" Make a list of all the places students looked. Which places had pillbugs? Which places didn't? You will want to save this list for use at the end of the project when students return the pillbugs to the places where they were found.

Tomorrow they will be examining the pillbugs they have collected and placed in the pillbug hideaway. It will be a one-on-one indoor activity, quite different from today's exuberant group hunt.

Teacher Narrative

This is the Pillbug Project, and you are part of the project. It will last about an hour each day for ten days. You are going to do a lot of watching of an animal you may already be familiar with—the pillbug. (They have other nicknames, but let's not worry about that right now.) To some people, they look like little compact cars just cruising around. You are going to experiment with them, play with them, study them, and listen to a story about one called Patricia. Plan to read a lot, plan to laugh, plan to think—these are all part of the Pillbug Project.

The Pillbug Project was written so you could learn about pillbugs and have fun learning about them. The Pillbug Project is going to help you discover how to learn directly from things in nature. Once you know how to learn directly from things in nature, you can call yourselves scientists.

Later on in the project we are going to do a variety of experiments using pillbugs, but first you have to meet some pillbugs and get to know them better. And the best way to do that is to get some. So you're all going to go outside and look for them. Where will you find them? I'll give you some hints. They hide. They hide from the Sun. They hide from things that like to eat them. They hide from dry air. They hide from the heat. The best place to hide from all of these things is under—under rocks, under boards, under flower pots, under a log, under a rotten old newspaper someone threw in the weeds.

There is another good place to hide from all these things that pillbugs have to hide from—in a miniature jungle. Imagine that you are a pillbug. What would it be like to walk on the lawn? It would probably be like a jungle, wouldn't it? You would be down on the dirt with tall, cool, grass blades towering way above you. You'd be out of sight of birds and frogs that like to eat you, and you'd be protected from the hot, drying sun. Imagine walking under a tree where there is a thick pile of pine needles. Crawling through the middle of the pile must be fantastic—the smell of needles, the dim light all brown in color, the stack of moist needles, each one seeming like a fallen tree compared to how big you are. Yes, if you were a pillbug, you could find a lot of places to hide.

Read the handout that you are going to get. It will be the first Research Note page of a lab notebook that you will make. The handout has questions on it that will start you thinking about what pillbugs are like. Then, go outside and find them. Search with your group, but each group should explore different places. Take with you the first page of your notebook. Also, take along a pencil and a plastic spoon (for gently scooping up pillbugs). When you are outside, follow the instructions on the page and see if you can discover answers for the questions.

Where Do Pillbugs Hang Out?

Materials for each scientist

a paper cup
a plastic spoon

What to do

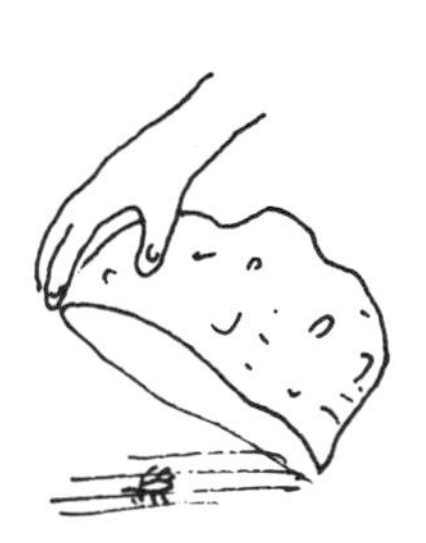

1 Spend 10 minutes looking for pillbugs. Spread out—each group look in a different place: under trees, in an open field, by the side of the road, in the lawn, close to planted flowers. Lift up things, look in things, look between things.

If you turn over a rock or a board, *carefully* put it back—you have opened up the home of many small animals and they need it kept just the way you found it.

2 After hunting 10 minutes, did you find any pillbugs? Not everything you pick up is going to have pillbugs under it.

What makes a good pillbug hiding place?

3 Read the questions below, and then look under enough different things with your group so that you feel you can answer the following questions.

Are there more, less, or about the same number of pillbugs under thin things like leaves and paper, as under thick things like boards and rocks?

Are there more, less, or about the same number of pillbugs under logs and wood as you find under rocks and bricks and stuff like that?

Are there more, less, or about the same number of pillbugs under things out in open, sunny places as you find under things in places where there is some shade, like under trees and bushes?

Are there more, less, or about the same number of pillbugs in dry soil as in moist soil?

Do all the pillbugs found in one place look the same?

Which seems a better place for pillbugs to be—in leaves or pine needles, under a tree, or out in an open lawn?

4 Now, each person collect about five pillbugs in a paper cup and take them back to the classroom. Put them in the pillbug hideaway.

Be careful anytime you pick up a pillbug or any other small animal. They are delicate and you can easily hurt them. Gently nudge them into a spoon or onto a piece of paper instead of trying to pick them up with your fingers.

5 Look around outside when you go home tonight. See how good you are at guessing where pillbugs hang out.

Second Day

Teacher Note

This is a day for observing and honing powers of observation. Spoon a few pillbugs out of the pillbug hideaway into a paper cup for each group of students. The groups (or pairs) of students can then pour pillbugs onto paper or their hands to observe them. Encourage the students to really look at the pillbugs, to watch their behavior carefully.

Try to make sure that each student gets involved with this activity. Pillbugs are nice to work with. They don't bite, aren't slimy, don't slither, and are cute (to some people, anyway). If those hesitant to touch bugs can learn first with a benign pillbug, they may gradually widen their zone of tolerance for "creepy, crawly things."

Questions you may want to ask as students examine their pillbugs include "Do all pillbugs have the same number of legs?" "Do all pillbugs have the same number of antennae? Segments?" "Where is the mouth of a pillbug?"

Should some students finish their observations before the others, you might encourage them to do a careful sketch of a pillbug. Arm them with a hand lens, and tell them to make sure that things like the legs and antennae are drawn realistically, perhaps even have them sketch both Hikers and Rollers and indicate differences in appearance between the two types.

You may discover that students would like to explore other aspects of pillbugs beyond the ones on this day's handout. If students suggest original research ideas, a class discussion on conditions of research is a good idea: Research must contribute to answering a question and research must never hurt the pillbug.

Materials List

- Pillbugs—The students need something to look at.
- Magnifying glass or hand lens—This isn't essential, but it does make it a little easier to see details.
- A rough surface like a rug or loose dirt—Anything a pillbug might be able to grab onto to roll itself back on to its feet.
- A cup of water—This is to quickly dunk the pillbugs in to see how they shed water.
- Two pieces of paper—One piece is for wet pillbugs to walk on so students can see their trail. The other piece is for Explorations #3 and #4 where the pillbugs will be walking on the paper held at various angles.
- A smoother or rougher surface than paper—So students can repeat Explorations #3 and #4, just for comparison.
- Post-it™ notes—Each student will need one of these for Exploration #4, when the class makes a bar graph.

Teacher Note

For Exploration 4, you have one preparation task. On the chalkboard or on newsprint paper draw a number line ranging from 0 to 25 cm. After students have measured their pillbug's path, each team will put a Post-it™ note above the number of centimeters their pillbug traveled in 15 seconds. When complete, the class will have generated a bar graph. They should be able to estimate the average distance a pillbug crawls in 15 seconds by just looking at the distribution of Post-it notes on the graph. You may want to explore other information that students can see from the graphical representation of their data, and later look at other bar graphs that compare with the one students created.

Tell students not to be concerned about the path the pillbugs take during the 15 seconds. The pillbugs may crawl in circles, but what is of interest in this investigation is determining how far they crawl.

To help students' extend their thinking you could pose some discussion questions. "Why might different pillbugs crawl at different speeds?" "When might it be helpful for a pillbug to be fast? When might it be harmful?"

Teacher Narrative

Let's get back to the Pillbug Project. Later today you are going to do some investigations to find out about pillbugs, but first you need to know a little about pillbugs.

I have a bunch of books at home. Some of them are called *reference* books. You can use reference books to look up things you want to find out about. Libraries also have a lot of reference books. I have looked in many books on plants and animals, but almost none of them say anything at all about pillbugs, and the rest that do say anything about pillbugs don't say much.

I think I know two reasons why it is hard to find information about pillbugs. First, books are usually written about groups of things—birds, mammals, trees, fish—and pillbugs aren't in any common group. So our books on these groups of animals don't mention pillbugs. Second, pillbugs don't harm people or domestic animals or crops, and they aren't economically important. So not much attention has been paid to them by scientists.

Oh well...maybe it's better this way. Since not many people know much about pillbugs, you might discover something no one else knows about them. Let's discuss what kind of things we know about pillbugs already—and I'll give you a little bit of information from the reference books.

1. They are not insects. Insects have only six legs. These guys have more. How many legs do pillbugs have? How many segments do they have? How many antennae do they have?
2. They belong to a group of animals called *crustacea*. Crabs, shrimp, and crayfish are other animals in this group. Most crustaceans live in water. The pillbugs you will

be studying are about the only type of crustacean that have figured out how to live their entire lives on land.

3. Though we are calling them pillbugs in this project, not everyone does. People in England call them "wood lice." Many people call only some kinds of pillbugs "pillbugs" and call other ones "sow bugs." Some people call them "land isopods." We'll call them "pillbugs." The Pillbug Project sounds better than the Land Isopod Project. But there is still a problem. Some books have the word "pillbug" with a space between the *pill* and the *bug*, but others have *pill* and *bug* together as one word. I like the one-word way. Pillbugs are too small to need two words.
4. Crustacea and insects are a lot alike. But there are some important differences. Both groups of animals have a hard outside and no bones on the inside. But insects have a coat of wax on the outside. This acts like an inside-out raincoat; instead of keeping moisture outside, it keeps them moist on the inside. If you scrape off the wax and let an insect walk around on a dry day, the moisture in its body will evaporate, and it will dry out and die. In fact, that is how some kinds of flea powder work. If you look at flea powder under a microscope, you will see it has sharp edges. These edges scrape on the flea when it walks around and soon rubs holes in its wax, and the flea dries out and dies. Most crustacea don't have this wax coat. So if a crustacean walks around on a dry day, it will dry out and die just like the flea with holes in its wax coating. Why do you think pillbugs try to be under things and out of the sun?
5. The hard outside covering that a pillbug has instead of bones cannot stretch or grow. So, in order for a pillbug to grow, it must crack open its coat and crawl out of it. The process is called *molting*. Under the old coat is a soft, new one. Once the pillbug is out of its old coat, the new coat expands, and then chemicals from the pillbug's body are added to it to make it hard. The pillbug then has room to grow within its new coat. The pillbug's coat covers all of its outside—every tiny leg, its antennae, everything.

   So molting is a complex job—for you, it would be like crawling out of hard pajamas that completely covered you. The pillbug crawls out of the back half first, and then a few days later, it crawls out of the front half. Keep your eyes open for a pillbug with a front half that looks different from the back half. That one will be half-way through with molting. Keep it in a container, and look at it from time to time for a day or so. Be extra careful with a pillbug that is molting. Molting is a difficult process, and animals are more delicate while they are undergoing it. If you're lucky, you may be able to watch a pillbug crawl out of the front half of its old coat.

Can you think of any other animals that molt? What special needs might a molting animal have?

6. There may be several kinds of pillbugs around where you are. Some places have as many as eight kinds. Some of those kinds can roll up into little balls when you touch them. We'll call all of those kinds ROLLERS. The rest of the kinds just keep on hiking when you touch them. We'll call them HIKERS. Some of you might want to collect as many different kinds as possible. Even deciding how many kinds you have is a challenge.

There are a lot more things to know about pillbugs, but now it's time for you to discover a few things on your own.

Today's Research Notes has things for you to do with pillbugs. There are questions for you to answer and explorations for you to do. Try to read everything yourself and do things without asking a bunch of questions. If you really do have a problem that you can't figure out by asking another student or reading the handout again, do ask. But I bet you can do almost all of it without help.

Today it is important that you start with the first observation in the Research Notes and do it carefully before doing further explorations. Each observation will help you get to know pillbugs a bit better. You, like any scientist, will be writing down what you find out, using your Research Notes as your scientific notebook.

Teacher Narrative

Please clean up everything and put the pillbugs back in the pillbug hideaway. Tonight, if you can find pillbugs around your own neighborhood, you may want to look for other things that pillbugs can and cannot do. You might see whether they can crawl on flour. Can they crawl upside down on cardboard or paper? Get a bunch and look at them carefully. Sometimes you will find one with only one antenna. Let it walk on the floor. Does it crawl as straight as ones with two antennae? Tomorrow, tell us what you found out.

What Are Pillbugs Like?

Materials for each team of scientists

pillbugs from the hideaway
a rough surface: rug, or loose dirt
water in a cup
a piece of paper marked off in 1 cm squares
a plastic spoon
a watch with a second hand or a digital stopwatch
a Post-it note

What to do

1 Work with your teammates on these questions and observations. Go to the pillbug hideaway, and using a spoon, gently pick up a pillbug and bring it back to your desk.

2 Let the pillbug walk around while you watch it carefully. Let it crawl on your hand.

Where are its eyes?

How many legs does it have?

If you touch it, what does it do?

Would you call it a Roller or a Hiker?

See the two wiggly "whiskers" in the front? They are called antennae. What do you think pillbugs use them for?

Exploration #1—Can pillbugs turn over?

What to do

1 Some pillbugs can't turn over. Turn a few Rollers and a few Hikers gently on their backs on top of something smooth and hard like your desk.

2 Watch them for a couple of minutes.

Can the Rollers turn over? How do they try?

Can the Hikers turn over? What do they do?

3 Now gently turn them on their backs on something rough like a piece of rug or loose dirt and see whether they can turn over.

Can the Rollers turn over? How do they try?

Can the Hikers turn over? What do they do?

Exploration #2—What do pillbugs do if they get very wet?

What to do

1 Dunk a pillbug under water for two to three seconds then take it out, put it on a piece of paper on your desk, and watch it. Watch one that is dry, also.

2 Write down anything the wet one does that is different from what the dry one does.

Why do you think the wet one behaves as it does?

Exploration #3—What do pillbugs do if they come to an edge?

What to do

1 Have one person on your team hold a piece of paper flat and a couple of centimeters above a desk.

2 Someone else on the team should now carefully pick up a pillbug with a spoon and put it on the paper. If it needs help getting on its feet, help it.

3 Watch the pillbug carefully when it walks to the edge of the paper. If the pillbug falls off, pick it up and put it back on the paper.

Is the pillbug a Roller or a Hiker?

Describe what it does when it gets to the edge.

Can it crawl over the edge and walk on the bottom side of the paper?

Exploration # 4—How fast can a pillbug move?

What to do

1 With a plastic spoon, carefully place a pillbug in the center of the paper marked with 1 cm squares. Write an "X" in the square with the pillbug.

2 When the pillbug starts crawling, note where the second hand is on the watch. After 15 seconds, mark the square where the pillbug has crawled to. (If the pillbug has rolled up into a ball, start again with another pillbug.)

3 Carry your pillbug back to the pillbug hideaway.

4 Now, count the number of squares in between the two X's, which will be the number of centimeters the pillbug crawled. Don't count the square where the pillbug started, but do count the one it was in at the end of 15 seconds.

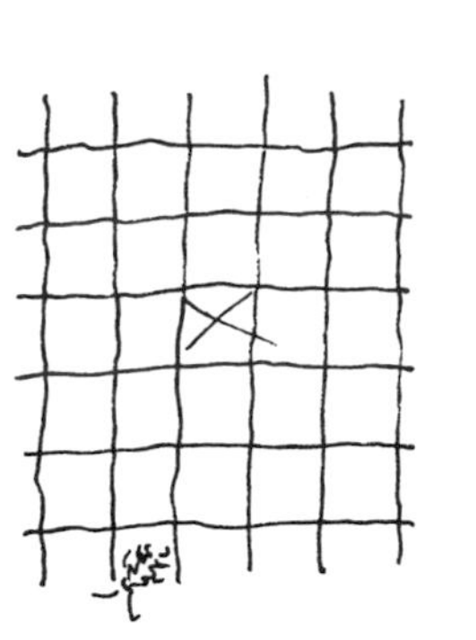

5 When your teacher tells you, have one team member put the Post-it note on the number line above the number of centimeters your pillbug crawled in 15 seconds. If there is already a piece of paper on that number, just put your paper directly above it, so it looks like a column of paper.

When all the teams have put their pieces of paper on the number line, then you will have a graph of the distances everyone's pillbugs crawled in 15 seconds.

How many centimeters did your pillbug crawl in 15 seconds?

How many centimeters do you think your pillbug could crawl in a minute?

Was your pillbug moving at the same speed all the time?

Were the starting and stopping points the same distance apart for all the pill-bugs?

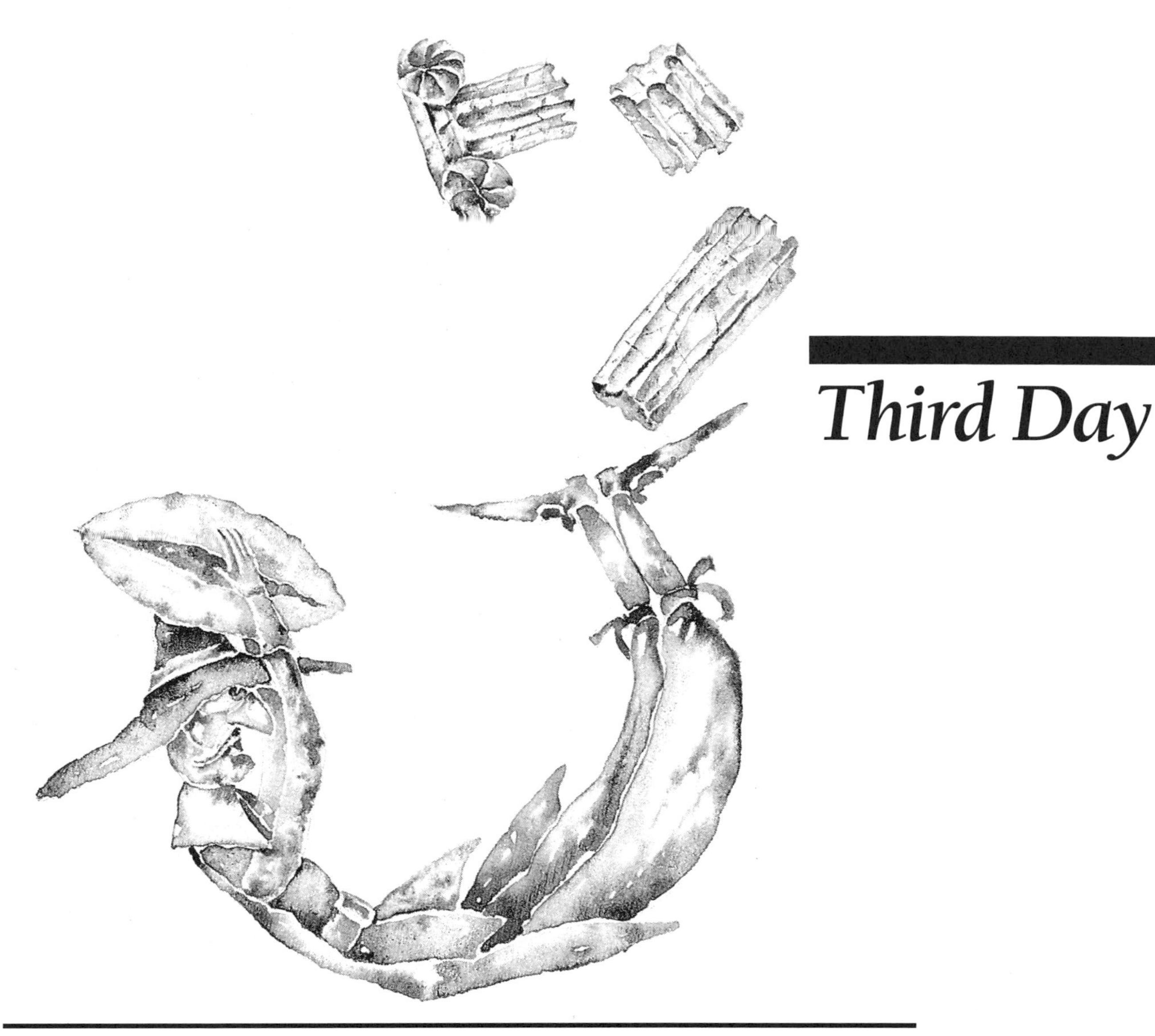

Third Day

Teacher Note

Patricia Pillbug is a fun story in three installments. Though it has a dab of biological information and a hint of a moral tossed in, you should read it as fun.
Following the story, the students will set up observation stations outside to determine where pillbugs are. Read over the handouts ahead of time and decide how to divide the class into the three groups: the PILLows, PILLars, and PILgrims.

Materials List

For the PILLows...

- Smooth-sided plastic glasses—These are for the pitfall traps the PILLows are going to set out.
- Some wet dirt and grass—These serve as "pillows" to cover the bottom of the pitfall trap and keep the overnight guests moist and happy.
- A spoon or a gardener's trowel—Used for general digging purposes.

For the PILLars...

- Bricks for pillbugs to hide under—The students will choose from the list on p. 39 places they want to test these brick "pillars."
- A spoon or small shovel
- A sink or bucket—Something that can be filled with water to soak the bricks in.
- A permanent marker—The PILgrims will also need one of these, for labeling purposes.

For the PILgrims...

- Objects for the pillbug "pilgrims" to hide under—Look at the PILgrims' list on p. 43. Depending on what your students choose to test, you'll need anything from rotten wood to wet newspapers.
- Labels—Maybe cards or pieces of paper to attach to the various objects.
- Glue, tape, a rubber band—Something to attach labels to their objects, depending on what needs to be attached to what.

Teacher Note

You will need to divide the class into three groups. Students within the groups will work in teams to set up their investigations. Their Research Notes pages have suggestions for their investigations and space for them to decide, on their own, testing locations and procedures. Encourage your budding scientists to come up with their own ideas and then test them.

For the best results, don't do these experiments if it is going to rain within the next three days. It would be particularly devastating to pillbugs in the pitfall traps. They don't mind rain, but they can't live in water like their crustacean kin.

After the students have completed their three days of testing, they will have time to meet as a group to discuss and make a chart of their group's results. The students have space in their data tables to add the results of three other teams.

To add math concepts, you could guide students in making more bar graphs of their pillbug results, using Post-it notes or student-drawn pillbugs for the bars. To add social studies concepts you could help students create a map of the school yard, showing the various locations of their pillbug tests.

Teacher Note

It is very important that each pitfall trap be placed so that the lip of the glass is exactly flush with the ground. If the lip is raised, the pillbug will not climb up and over it to fall in. You may want to check on the PILLows group setting up the pitfall traps and see that they are doing it correctly. Also make sure that there is enough of a "pillow" of moist dirt or grass in the bottom of the traps, so that the animals won't dry out. But be sure that something doesn't provide a ladder for the pillbugs to escape.

Encourage the students who are putting out the pitfall traps and the bricks to try a wide variety of locations for their observation stations. But for the PILgrims who have different objects, make sure they keep all the objects in a similar area, an area they have reason to believe has a large number of pillbugs roaming around.

One final point—don't forget to stress that *not* finding pillbugs under some things or in some traps is still valuable information. Their exploration is not a failure because they don't find any pillbugs.

PATRICIA PILLBUG

(a just-pretend story)

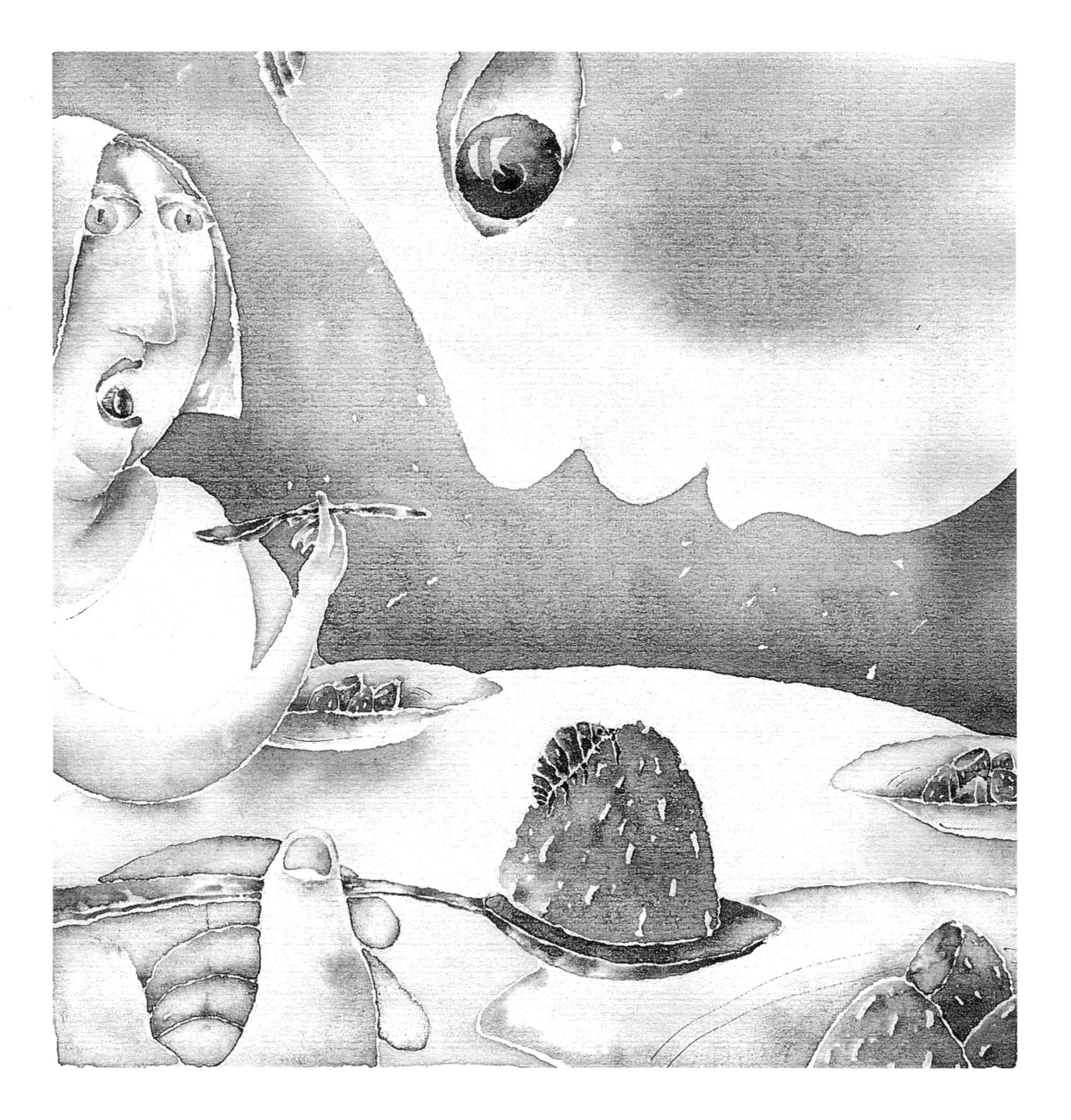

MEET PATRICIA PILLBUG. Patricia is just a normal pillbug, not much different from any other pillbug. She is named Patricia because it's easier to talk about things that have names. None of the other pillbugs in the strawberry patch where she was born know her as Patricia. Pillbugs don't name each other.

Patricia was born, as I said, in a strawberry patch. But *when* she was born is a bit harder to say. On September 6 the egg she was in, along with 27 others, emerged from inside of her mother and took up residence in a pouch between her mother's back legs. After four weeks, Patricia hatched out of her egg. You might say she was born then—or you might not—because she stayed within the pouch, bathed in liquid, for another few days at which time she crawled out and away and never saw her mother again. You perhaps could say that the day she left the moist brood pouch and first and last encountered the eyes of her mother was the day she was born.

Anyway, for eight months Patricia lived in the strawberry patch, spending the winter dining on decaying leaves. In late spring, the aroma of ripening strawberries became irresistible. It was time for fresh fruit. Patricia's approach to strawberry feasting was not dramatic. She started by crawling under a nice, ripe strawberry. Then she arched her back and started chewing. After an hour of eating, she had made only a tiny hole. After a couple of days, she had carved out a pillbug-sized cave. So she crawled in the cave, made it her home, and kept on eating, surrounded by strawberry smell, strawberry flesh, and strawberry juice. She must have been in pillbug heaven, nestled nicely within a combination bed, house, and meal. There she was safe from predators, safe from the drying sun, safe from everything—everything but...

"You pick that side of the row, Sarah, and I'll pick this side."

"OK, Mom. Wow! The berries are really big and sweet this year."

Sarah and her mother worked their way down the row getting closer and closer to where Patricia was nestled in her berry. When Sarah stepped quite close, Patricia's antennae picked up the vibration from Sarah's shoe hitting the ground. When Sarah reached out and picked Patricia's strawberry, Patricia extended her legs, wedging herself in her homemade cave. Her world tumbled, twirled, and lurched as her strawberry was lifted in the air and aimed right toward Sarah's mouth.

"Hey, Sarah-bum, put some of the berries in the bowl—we want them for dessert tonight."

"Oh, OK, I'll put *some* in," quipped Sarah as she dropped the berry, Patricia and all, into the bowl with what sounded like a small thump to Sarah but which must have really rattled Patricia's mandibles. Mandibles, by the way, are the mouth parts used by pillbugs to bite off tidbits of food. And there, within her strawberry, Patricia stayed as Sarah and her mother continued talking and picking and picking and talking until their bowls were filled. Then they carried the strawberries to the kitchen sink where they were dumped into a large strainer and rinsed under the faucet.

Great quantities of frothy water cascaded around Patricia's berry dousing Patricia and almost causing her to abandon her berry. Hey, a little water is fine for a pillbug, needed in fact—but who needs to be out in the middle of Niagara Falls? But she held on, and continued to hold on as Mom picked up berry after berry and cut their crown-of-green off with a sharp knife and then sliced them in half and dropped them in a glass bowl. By chance Patricia and her home-sweet-home were on the bottom side of the strawberry, so Mom didn't notice either one when she picked up that strawberry. *Slash*—the knife severed the top of the berry; *schliiick*—it cleaved the berry in half, both times missing Patricia by a scant tenth-of-a-millimeter. If she could have blinked, Patricia probably would have, but pillbugs have no eyelids. If she could have screamed, she might have, but pillbugs have no voices. So she just held on, and even that is hard for a pillbug to do.

ROM WAY ABOVE her came a hailstorm of white particles as Mom sprinkled sugar on the bowl of berries. Then suddenly chaos. Patricia was bowled and rolled, tossed and tumbled as Mom gave the bowl a quick stir. Then all was quiet, quiet except for the call to dinner and the dinner -time conversations about school, play practice, and weekend plans, none of which a pillbug can begin to understand, and which aren't particularly important to a pillbug being served up for desert.

"Yahoo! Strawberries for desert," yelped younger brother Ben.

"Hold your bowl over here, Ben."

"Is that all I get?"

"OK, OK. One more spoonful," Mom said as she dipped the serving spoon into the glass bowl for a fourth time, this time scooping up Patricia's very berry complete with a by-now sticky Patricia.

Patricia held very still as Ben's spoon dove into his bowl, scooping up the berries all around hers. On the fifth trip from mouth to bowl, the spoon nestled under Patricia's berry, and Patricia started her voyage straight toward the open cavern of Ben's mouth. Closer and closer and...

Suddenly Ben's mouth closed and then opened long enough to let out one word—"Yuck!" Two brown eyes stared right at Patricia's strawberry, right at her cave, right at her.

"Yuck?" questioned Mom.

"There's a bug in my strawberry."

"Well, spit it out."

"It's not in, so I can't spit it out. Here, look at it." Ben extended his spoon to his mom's face.

She tilted her head back, as parents tend to do when kids hold things close for them to examine, adjusted her glasses, and then announced, "It's one of those roly-polies. I see a lot of those bugs in the strawberry patch."

"Let's see it," chimed in Sarah. "Oh…that's a pillbug. It's not really a bug you know, it's more like a shrimp or a crab. We're studying them at school right now."

"Double yuck! Crabs in my strawberries," announced Ben.

"They're not crabs, they're isopods. But they're related to crabs, like mice are to bunnies. We're learning a lot about pillbugs," Sarah said proudly.

"Oh? Did you learn how to kill them? I don't want them in my strawberries," said Mom.

"Aw, Mom. They usually don't hurt anything. They are really kind of nice."

"Well, why don't you be kind of nice and carry out the garbage while Ben clears the table."

Sarah, being in a good mood, jumped up from the table and immediately took out the trash. Since she was outside cramming things into the overstuffed garbage can, Sarah didn't hear Ben's next question or her mother's reply. If she had, this story might have been different.

"What shall I do with the bug in the strawberry?" asked Ben.

"Well, put it on your plate, and then dump it in the garbage disposal after dinner," said Mom.

"Prepare the landing pad for the isopod," Ben said in a deep voice as he dramatically flew his spoon down to his dinner plate and deposited Patricia and berry in the middle of his uneaten creamed corn.

Teacher Narrative

And here we will have to wait until tomorrow for more of the story. We will have to wait because now it is time to set up some observation stations.

Today you are going to work in your three groups—the PILLows, the PILLars, and the PILgrims. Each group will set up observation stations to find out something about pillbugs.

The PILLows are going to do an investigation to find out where pillbugs walk around. It might be that there are places where there are many pillbugs, but they stay hidden, and other places where there aren't as many, but these pillbugs do a lot of hiking around. The PILLows are going to set up traps to catch the ones that are out walking.

The PILLars are going to do an investigation to find out where pillbugs like to hide. When you looked for them a few days ago, you found them where they were hiding, but it might be that they would rather hide somewhere else if they had a choice. So the PILLars are going to put out brick "pillars" in different places and orientations to see what places attract the most pillbugs.

The final group, the PILgrims, are going to try to find out what pillbug "pilgrims" prefer to hide under. You found them under all types of things on our first day of the project, but all those things were in different types of locations—some dry, some moist, some shady, some bright. So the PILgrims will put out different objects, all in the same general environment, and then they will see which objects the pillbug pilgrims prefer to hide under.

You will check your observation stations each day for several days, and on the last day your team will give a report to the class. Not all of you will find pillbugs under your objects or in your traps, but not finding pillbugs gives you information too. So don't be discouraged if others in your group get a lot of pillbugs, and you don't. That's just how science works out sometimes.

The PILLows—Where Do Pillbugs Walk?

Materials list for each team of scientists

a small shovel or spoon
plastic cups

What to do

1 You PILLows are going to use plastic cups as traps to discover places where pillbugs walk around a lot.

Work in teams of two. From the list below, select a place to put your traps. (Each team of scientists should pick a different place.)

Here is a list of some of the places where you might put your traps.

- in the lawn where it is always shady
- in the lawn where it can be very sunny
- in tall grass or weeds
- at the edge of a dirt path or dirt road
- next to planted flowers or bushes
- at the base of a large tree
- in a drainage ditch
- right next to a wall, so that the trap will be in the path of a pillbug walking along the base of the wall
- any place that you think will have a bunch of pillbugs walking around

2 Each team should go to the place it selected and carefully dig a hole with a small shovel or a spoon. The hole should be just the size and shape of the cup, so that when you put the cup in the hole, the lip of the cup is just even with the ground.

3 Put the cup in the hole, and smooth the dirt right up to the edge. If a pillbug wanders up to the edge, it will tumble in and not be able to get out.

A trap that catches things that fall into it is called a *pitfall trap*. A big pitfall trap can catch an elephant.

4 Put a "pillow" of wet dirt or grass in the bottom so that all the trapped pillbugs will be moist and happy during their overnight stay. But leave most of the trap empty so that the pillbugs can't use their pillow to climb out.

Where did you put your trap?

5 Check the number of pillbugs in your trap each day for the next three days. After you check your trap, release any overnight visitors several meters away from your trap and replace the moist grass or dirt with a fresh, new "pillow."

6 Record each day's results in the data table below. You will notice that the data table has space for data from other trap locations. At the end of this experiment, you will add the numbers from the other PILLow teams.

Trap location				
Number after first day				
Number after second day				
Number after third day				

7 After three days, meet with the other PILLow teams and fill in the rest of your data table with the results from the other PILLow teams.

Then, as a group, answer the questions below to help get ready for the PILLow group's report.

Questions for your group report

Was there anything special about your team's pitfall trap?

What other organisms did you collect in your trap?

If you did collect other organisms, could this have affected your number of pillbugs?

Which location of all those tested by the PILLow teams seemed to have the most pillbugs?

What do you think attracted the pillbugs to this area?

Which location had the fewest pillbugs?

What do you think you could do to attract pillbugs to this area?

What else do you feel you can conclude from your group's investigation?

The PILLars—Where Do Pillbugs Hide?

Materials list for each team of scientists

a small shovel or spoon
a bucket of water
a brick
a permanent marker

What to do

1 You PILLars will test different places to see where pillbugs prefer to hide—where they like to be, and where they don't like to be. All the teams will use the same type of object but will put them in different environments. So the pillbugs will choose a place for its particular features, not for the object that is there.

You will work in teams of two. First, label your brick, using a permanent marker.

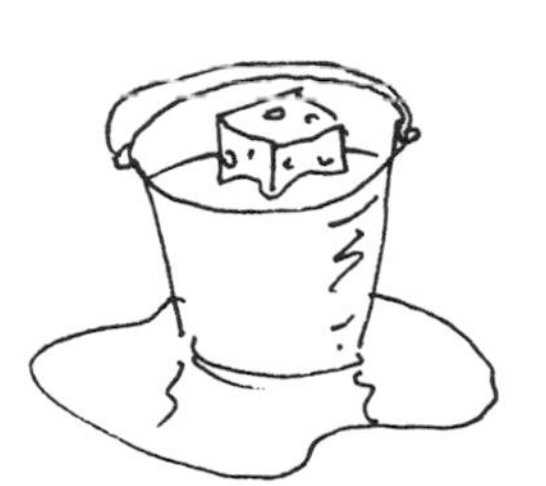

2 Soak the brick in the water for about five minutes.

3 Select a place to put your brick from the list below.

Here is a list of places where you might set your bricks.

- at the edge of a lawn where it is usually sunny
- at the edge of a lawn where it is usually shady
- somewhere on dirt where there are bushes or flowers nearby
- in the dirt in a field where there are a lot of weeds and tall grass
- at the bottom of a ditch
- near or touching the wall of the school on the north side, where the sun never hits (try to put them on dirt near grass or weeds)
- near or touching the wall of the school on the south side, where the sun hits directly (try to put them on dirt near grass or weeds)
- any other good places your team agrees on

4 Find a flat area on the ground. Move away little rocks and sticks if that will help the brick lay closer to the ground. Put the brick down and push it slightly into the ground.

5 Every day for the next few days go and inspect the area under the brick "pillars" by gently lifting them up and counting the number of pillbugs underneath. Count how many of the pillbugs are Rollers and how many are Hikers. Write down the results in your data table. You will notice that the data table has space for data from other locations. At the end of this experiment, you will add the numbers from the other PILLar teams.

PILLar Observations:

First day of investigation

Has the weather been cloudy or sunny?

What has the temperature been outside?

Has it rained?

Is there anything special about where your brick was placed?

Second day of investigation

Has the weather been cloudy or sunny?

What has the temperature been outside?

Has it rained?

Third day of investigation

Has the weather been cloudy or sunny?

What has the temperature been outside?

Has it rained?

Brick location								
	Hikers	Rollers	Hikers	Rollers	Hikers	Rollers	Hikers	Rollers
Number after first day								
Number after second day								
Number after third day								

6 After gathering data for three days, meet with the other teams of the PILLar group and add their data to your data table.

Then, as a group, answer the questions below. This will help you get ready for the PILLar group's report.

Questions for your group report

Which location seemed to attract the most pillbugs?

What do you think attracted the pillbugs to this area?

Which location had the fewest pillbugs?

Does one type of location attract more Hikers than Rollers? Why do you think this is so?

Did the weather seem to affect how many pillbugs were under the bricks?

What do you think you could do to attract pillbugs to this area?

What else can you conclude from your group's investigation?

The PILgrims—What Objects Do Pillbugs Prefer to Hide Under?

Materials for each team of scientists

objects for pillbugs to hide under
labels for the objects
glue or tape
metric ruler

What to do

1 You PILgrims will test different objects to discover which ones pillbug "pilgrims" prefer as hiding places. Your group will work in teams of two. Each team will test a different type of object, but all of the PILgrim teams will put their objects in the same type of area or environment. Things like the amount of sunshine and the type of ground should all be the same. This will give each object a "fair test." You will be able to say that the reason pillbugs choose an object was because of the object's characteristics, not where the object was placed.

Now, to begin this project, pick a type of object to test from the list below.

Here is a list of possible things pillbugs could hide under.

- stacks of wet cardboard (you will need to wet these again each day)
- scraps of wood, all approximately the same size
- stacks of wet newspapers—about two centimeters thick (put a rock on top so they won't blow away) (you will need to wet these again each day)
- stacks of dry newspapers—about two centimeters thick (put a rock on top so they won't blow away)
- a pile of weeds or other plant material that no one wants (when you check this object you will need to look under and *in* the pile)
- rocks or pieces of concrete
- rotten wood—small logs or boards
- big, flat pieces of sliced pumpkin, watermelon, or squash
- anything else you want to test

2 Label your object with your names and a note telling people not to touch it. Figure out how to attach the label (rubber band, tape, glue, depending on what type of object you have).

3 As a group, all the PILgrim teams should decide where to put their various objects. Maybe under a large tree where there are leaves or weeds, but not on the lawn which could be hurt by leaving things on it. Try to pick a place where few people go, so that no one will bother your experiment. If you know of a place where there are lots of pillbugs, you might try that place.

4 Go to the area you selected and find or make flat areas to put your team's object so it will lie close to the dirt.

5 Check your objects for three days. Each day count how many Hikers and Rollers you find under your object. Be sure to carefully put the object back in its original position. Write down the results in the data table. You will notice that the data table has space for data from other object locations. At the end of this experiment, you will add the numbers from the other PILgrim teams.

What types of object did your team test?

Where did the PILgrim group decide to place all the objects?

How big is your object ?

Is there anything special about where your team put its object?

Object location								
	Hikers	Rollers	Hikers	Rollers	Hikers	Rollers	Hikers	Rollers
Number after first day								
Number after second day								
Number after third day								

6 After you have gathered data for three days, meet with the other teams of the PILgrim group and add their data to your data table.

Now, as a group, answer the questions below. These questions will help you get ready for the PILgrim's group report.

Questions for the group report

Which object seemed to attract the most pillbugs?

What do you think attracted the pillbugs to this object?

Which object had the fewest pillbugs?

Does a certain type of object attract more Hikers than Rollers? Why do you think this is so?

What else can you conclude from your group's investigation?

Fourth Day

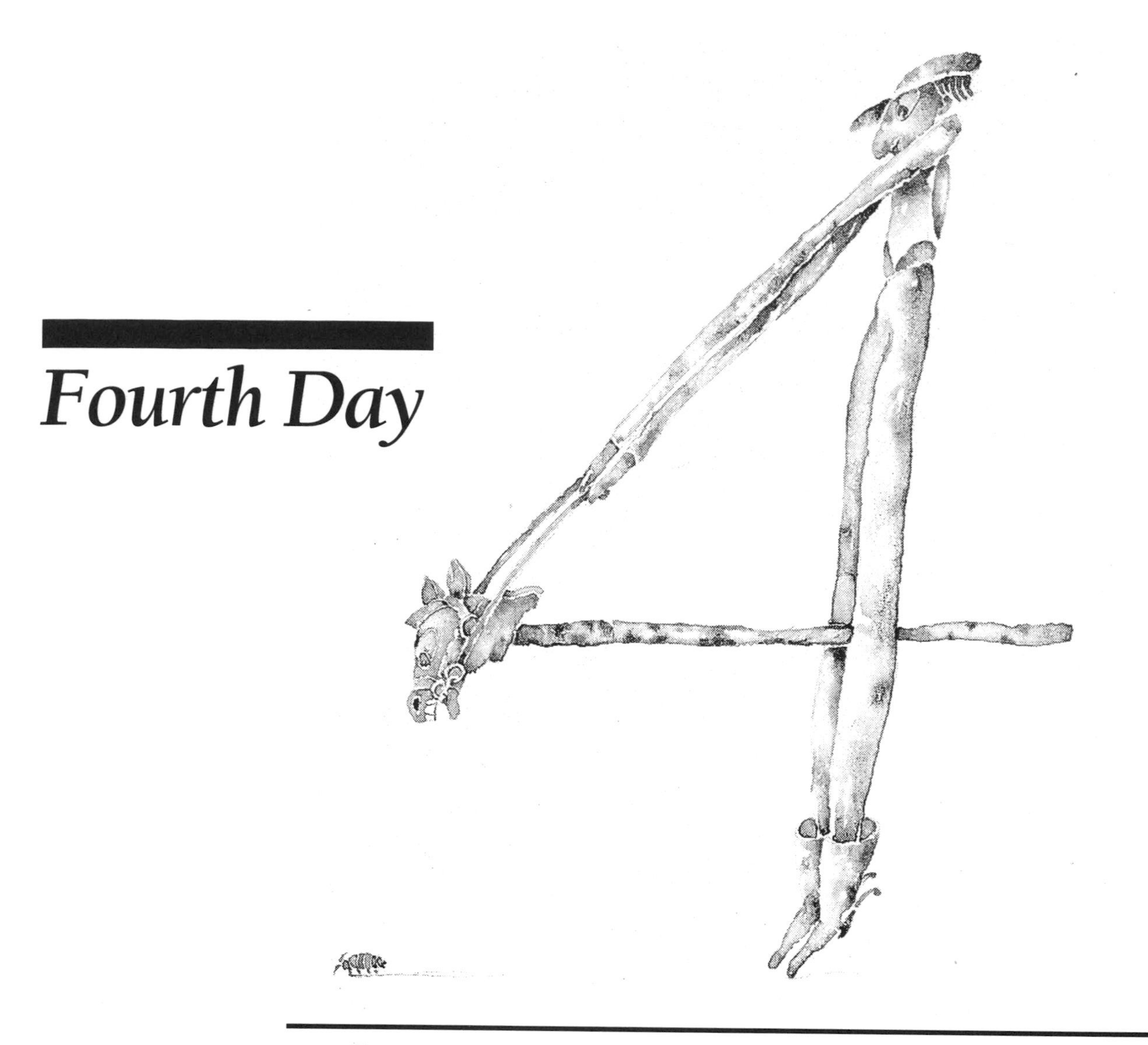

Teacher Note

Following the continuation of *Patricia Pillbug* the children are going to check their outdoor investigations and then it will be time for the pillbug races. Before the race you might encourage the students to make predictions about what characteristics of a race track, a pillbug, or the race conditions could make a race faster. There really is a great variation in the speed of individual pillbugs, and you could encourage students to test their predictions.

The races are fun but can verge on the chaotic. Because children can become concerned that the race track circles are not all equal, perhaps there should be an official measurer who makes sure all tracks are the same size. If the races are close, consider having a Finals Race with four or so of the fastest pillbugs on the course at the same time.

Enclosed is an official award which you can fill out and put on the bulletin board to honor the fastest pillbug. I suggest you gather the class after the races to present the award and have a brief discussion about what makes for a fast pillbug. Some questions you might consider are "Which would you expect to be faster—rollers or hikers?" "Do you think temperature affects their speed?" "Do you think the size of the pillbug affects its speed?"

You might even arrange a table of columns on the blackboard so that your students can write down their names, the times their pillbugs took to run the race, where they were collected, what kind they were (rollers or hikers), and how long (in millimeters) each pillbug is. Students can look at this table and try to find common features of the very fast or very slow pillbugs. This could also be a starting point for students to conduct more investigations on their own, at home or at school.

Materials List

- Sheets of brown wrapping paper at least 240 cm long—This is only if you decide to have the race track on paper instead of the ground. Tape two pieces together for the racetrack, particularly if it is too hot, too cold, or rainy on the day you hold the race. You'll also need tape if you choose this method of race track construction.
- Metric tape measure or meter stick—For measuring the race courses.
- A watch with a second hand or stopwatch—For timing the contestants.
- A piece of string and a piece of chalk—To draw the racetrack circles.

Teacher Note

The race course is just two circles, both having the same center. It can be indoors directly on a floor, on kraft paper (brown wrapping paper) on the floor, or outdoors—whichever is more convenient for you and the pillbugs. Remember that if the weather is too cool, pillbugs will be sluggish; and if the sun is too hot, 80 centimeters will be a long way for a small pillbug on a dry surface. If students are old enough, let them make the race courses. Or you may want to have an official race course circle-drawer or circle measurer—to ward off complaints about an unfair track.

There can be more than one race course. The pillbugs will be racing one at a time; each will take one to three minutes to complete the course. You may want to decide ahead of time how many courses you think is best.

The inside circle should be 20 centimeters in diameter, and the outside circle should be 180 centimeters across. All race courses should have the same surface, however, and each race course must be flat and the same size and shape as the others.

For a kraft paper race course, cut enough strips of paper to create a sheet at least 180 cm square. Tape the sheets together—the tape must extend the entire length of the paper. If you leave gaps, pillbugs might try to climb between the sheets. For a floor or outdoor pavement, be sure you can make a chalk mark on the surface to mark the track (or you might use tape on the floor).

Use a piece of chalk to mark an X on the ground or on the paper where you want the center of your circle to be. To make your circle, tie a piece of chalk to a string. With a pen, put a mark on the string 10 centimeters from the chalk. Hold the mark on the string on the center of the X that you marked on the ground. Draw a circle on the ground with the chalk by keeping the string tight and moving the chalk in a circle around the X. Now do the same thing again, only this time put a mark on the string 90 centimeters from the chalk, since the outer circle will have a 90 centimeter radius. Have one person hold that mark on the X in the center, while another pulls the string tight and draws the outside circle.

After the races, give students time to write their results in their research notebooks. Then bring them together to compare results, describe characteristics of fast and slow pillbugs, and give the award to the team with the winning pillbug.

More about Patricia Pillbug . . .

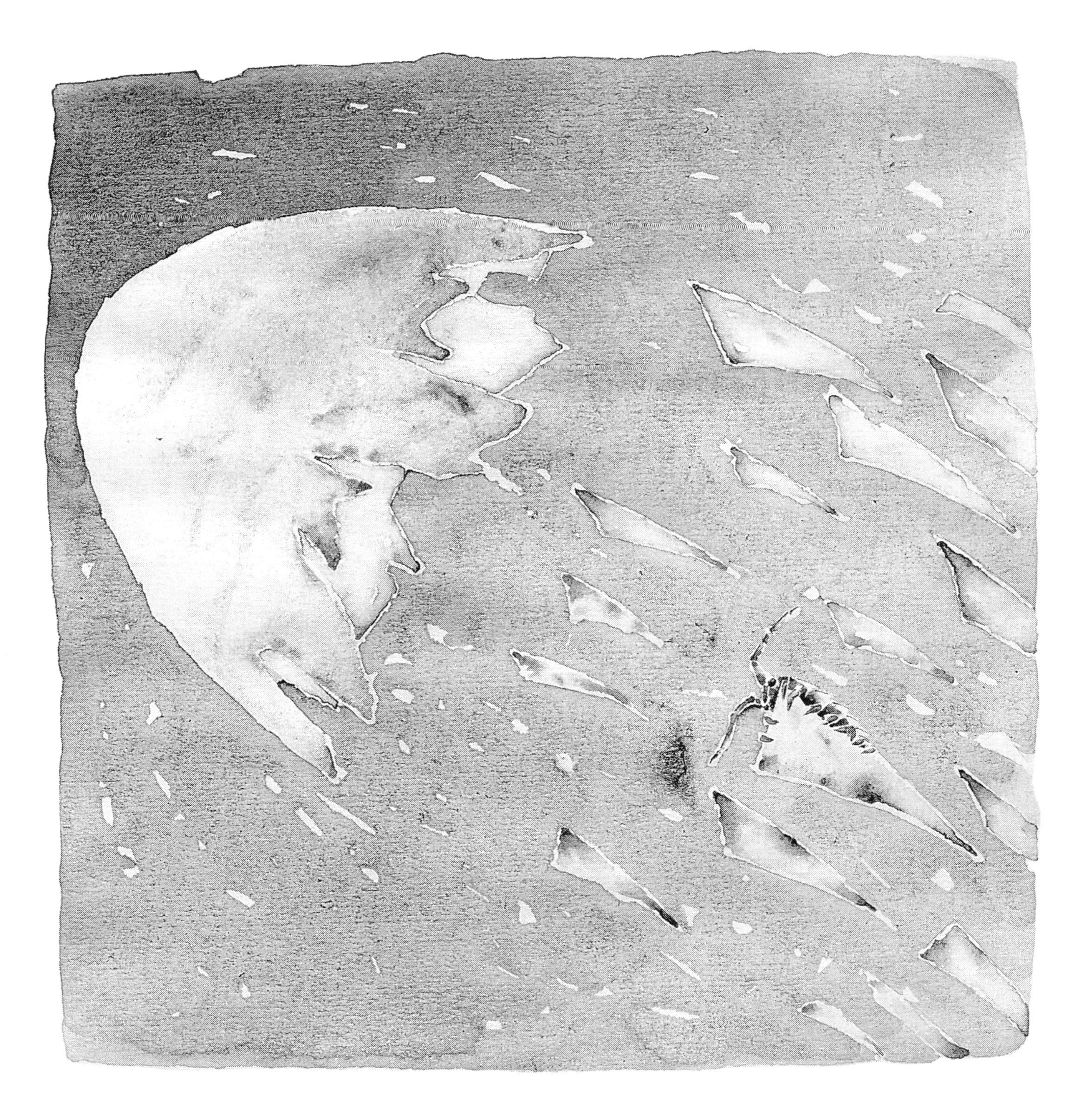

YESTERDAY THE STORY stopped just after Mom suggested that, after dinner, they would dump Ben's strawberry in the garbage disposal. A garbage disposal is a very loud machine in the bottom of some people's sinks. It grinds and chews and smashes things that are put in it. Now, it was a good thing for Patricia that she wasn't going into Ben's mouth since mouths often grind and chew and smash things that are put in them. But it wasn't such a good thing for Patricia that Mom wanted to put the strawberry with her in it into the garbage disposal. It doesn't matter *how* a pillbug is ground, chewed, and smashed: it all amounts to the same thing in the end.

Now, Patricia didn't know what was about to happen to her, but having her strawberry home partially submerged in creamed corn was not the most pleasant of habitats. She was about to take her first step to abandon her berry when Mom asked Ben to help clear the table. Suddenly the plate, Patricia and her strawberry, and the sea of creamed corn were airborne. Ben was pretending his plate was an airplane in the sort of desperate trouble that only jettisoning all the cargo would help. He flew the plate with the creamed corn and the strawberry with the hole with the pillbug with the name Patricia to the sink and, using a brush already covered with slimy mashed potatoes, coffee grounds, and chicken gravy, he scraped the strawberry into the garbage disposal, thereby saving the airplane plate from a fiery crash.

Into the deep chasm tumbled Patricia. You might think that being in a garbage disposal was a new horror to her. But no, because here was a place which was, in many ways, the kind of place where pillbugs hang out; it was wet, cool, and dark. If she could have, Patricia would have taken a big breath and sighed in relief—but pillbugs can't take big breaths. In fact, they can't take breaths at all. They breathe by letting in air through tiny holes somewhere around where their bellybuttons would be if they had bellybuttons, which they don't. The dark disposal got even wetter when Ben turned on the cold water.

When the water hit her, Patricia started moving. At the same instant, Ben started to move his hand to flip the switch on the wall that would turn on the motor that would start the bottom spinning that would fling the potatoes, gravy, coffee grounds, and the strawberry into spinning blades that would grind them all up so that they could be washed down the drain.

Patricia crawled as fast as her 14 legs could move her. She crawled out of the strawberry and up over an apple core as Ben's hand neared the switch. She slid down a banana peel as Ben's hand touched the switch, and she started to climb the wall of the garbage disposal as Ben pushed the switch up, turning on the machine.

H, THE ROAR! Things spun around. Patricia instinctively froze in her tracks as an eggshell smashed into the wall just above her head. A glob of mayonnaise splatted into her side and almost knocked her off. Onion skins whisked around like paper tossed in a storm. Patricia held on to the wall of the disposal while the twirling mass slowly got smaller as bits of it slid down the drain. Because of her poor eyesight, she never saw her wonderful, delicious-smelling, wet and cold, oh-so-tasty strawberry bed being flung right into the sharp blades of the garbage disposal. It burst and its red insides were shredded into juicy bits. Most of it got sucked right down the drain but one big blob bounced off a piece of carrot and flew right at Patricia. Patricia didn't see that wad of her wonderful bed as it was flung toward her, and she couldn't have moved fast enough to avoid it even if she could have seen it.

Splat! It smashed her right between her eyes, which is also right between her antennae. She smelled and tasted it with her antennae. It was delicious, but it wasn't a good time to eat part of her bed, particularly since a potato peel immediately followed the strawberry blob. The peel whisked her right off the wall.

At that exact second, the telephone rang. Patricia, whose hearing consisted of the ability to sense vibrations, couldn't sense the vibrations of the phone ringing over the tremendous trembling of the disposal. But nevertheless it was the most important phone call of her life, even though it was a wrong number. It was a man trying to call Piece o' Pizza to order a turnip pizza. People, you see, have trouble hearing anything on the phone when their little brothers are flying jet airplanes in the kitchen and when the garbage disposal is on.

So Sarah reached up and turned off the disposal, shouted at her brother to fly his plate to China, and picked up the phone. She did so just as Patricia, holding on to the same potato peel that knocked her off the wall, fell to the horrible bottom of the machine. The bottom was spinning wildly around, and Patricia tried to hold on to it as she was being spun outwards toward the shiny steel blades. But she couldn't. She slid out toward those blades as the machine slowed down. Closer and closer she got until, just as the floor stopped spinning, she touched the blades.

All was quiet except for a gurgle as the remains of her bed, and almost everything else that had been in the disposal, slid down the drain. In the distance Sarah was saying, "I am sorry, we don't have any turnips. Besides, turnip pizza sounds as yucky as eating a strawberry with a pillbug in it, and our oven isn't on so we can't cook you a pizza, and this isn't a pizza store because you have a wrong number, so good-bye."

Patricia would have just lain there panting, her teeth almost touching the blades of the disposal if only she could pant and if only she had teeth. She couldn't and didn't, so she started crawling—up, since that was the only way out unless you were first ground up, in which case down the drain was an option. So up she climbed. Up out of the disposal into the sink, only to appear in front of the eyes of Mom looking disgustedly down at her.

"Bugs! Bugs everywhere. First in a strawberry, now walking boldly around the sink," said Mom who had come to the sink to finish cleaning up after dinner.

At this point, Patricia did not put her hands on her hips, and declare, "Harumph, bugs indeed. And I certainly never asked to or wanted to walk around your sink. It just happens to be on my way out of here." She didn't do this because pillbugs don't have hips, and because pillbugs can't even say "Rudolph," let alone all of these words, and because she couldn't begin to understand what Mom said in the first place. So she kept on walking.

"Sarah, could you please take this bug outside, then clean your room, do your homework, wash the floors, and fold the laundry for me," asked her mother.

"Sure, Mom," answered Sarah, which proves that this is a just-pretend story since no kid in this world ever answered "Sure, Mom" to a mother who asked her to do all of that stuff.

Nevertheless, Sarah walked over to the sink and looked at Patricia who looked right back. "Mom, I bet that's the guy that was in Ben's strawberry. I wonder what kind it is," said Sarah.

"I thought you knew what kind of bug it was. Didn't you even say something about studying them at school?" asked Mom.

"Mom, they're not bugs, they're pillbugs. But there's more than one kind of pillbug. Some kinds we call Rollers and others we call Hikers. If you poke it and it rolls into a ball, it's a Roller," said Sarah.

"Which type is this one?" asked her mother.

"I'll test and find out," said Sarah, and she reached out and gave Patricia a gentle poke.

At least Sarah intended it to be gentle. But to Patricia, Sarah's finger was as big as a freight train is to you. Patricia skidded sideways and instinctively reacted as if this were an attack from a vicious predator, trudging off as rapidly as her stubby legs could trudge her.

"He's a Hiker," announced Sarah, proud of her observation. Sarah reached down with a spoon to pick her up and said, "I'm going to take him to school tomorrow."

Given a choice, Patricia would have picked being curled up in a strawberry over going to school, which shows that pillbugs and a lot of children are not always all that different.

"How do you know he is a he?" asked her mother.

"I don't. But I think he is a he."

Perhaps it's a good thing that Patricia couldn't hear what people were saying about her, calling her yucky, planning to dump her into a garbage disposal then planning to tote her to school, and now saying that she was a he. Patricia might have been upset if she had comprehended any of this. Even without understanding what was going on, Patricia knew she wasn't where she wanted to be, so she kept on running.

"He sure can run fast. I'm going to have fun with him tomorrow. I think I'll call him Fred," said Sarah as she reached down, picked up Patricia, and dropped her into a recently-emptied yogurt container.

If Patricia had been the human being doing the naming, she might have named Sarah "Snailslime" after all that had happened. It had been a rough day for Patricia and unbeknownst to her, her adventures were not yet over.

Teacher Narrative

We will see what happens to Patricia and Sarah tomorrow. Now, take 10 minutes—but no more that 10 minutes—to go outside and carefully check your observation stations. Remember to write down your results on your Research Notes. We'll meet back here in 10 minutes.

Teacher Narrative

It's almost time for the races!

What kind of pillbug is fastest? We are going to have a contest to find out. You will be divided into teams, and the team which finds the fastest pillbug will be the winner. Experiment with different pillbugs: try Rollers, try Hikers, try big ones and little ones, try wet ones and dry ones, try ones you just caught, try ones you have kept for a couple of days, and try ones caught in different places.

You are going to get a page containing some questions about your pillbugs. Record what you learn. Then we will share our observations so we can all learn something about fast and slow pillbugs.

Teacher Note

The text on the following page consists of the rules for the pillbug races. Read the rules aloud after the students come back from checking their observation stations. Make additional copies of the rules so you can post them for students to refer to.

Rules for the Pillbug Races

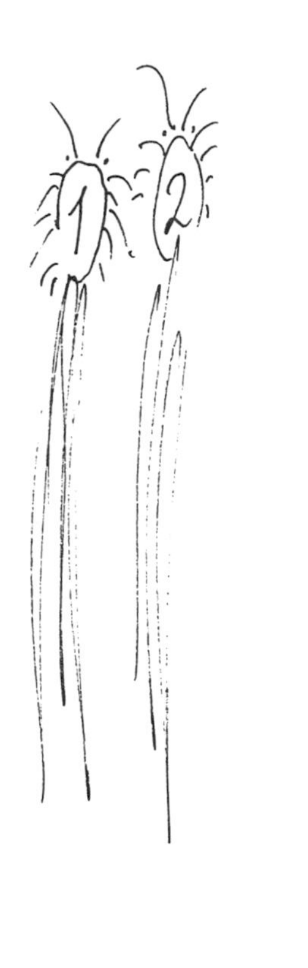

1 Each team of two students can enter two different pillbugs in the race.

2 Race one pillbug at a time. The time it takes for it to go from the inside circle to the outside circle is its score.

3 There will be an official timer appointed for each race course. When a pillbug crosses the inside circle, the timer will start timing and stop when the pillbug crosses the outside circle. Each team should write down how many seconds it takes its pillbug to run the race.

4 Each pillbug must finish in less than 3 minutes or its score does not count and the team may race an extra pillbug. But no team may race more than three pillbugs, even if two or more do not finish in less than three minutes.

5 While pillbugs are inside the inside circle, you may do anything that doesn't hurt them to start them racing. However, as soon as the pillbug leaves the inside circle, humans must stay outside of the outer circle. It is against the rules to poke, blow, or in any way move a pillbug during its race.

Time for the races—and may the best pillbug win!

We're off to the Races!

Questions about the contestants

Answer the following questions about your team's two pillbug contestants (Pillbug #1 and #2).

Where did you find your pillbugs?

Pillbug #1

Pillbug #2

How long did you keep them before the race?

Pillbug #1

Pillbug #2

Is it a Roller or a Hiker?

Pillbug #1

Pillbug #2

Carefully measure how many millimeters long it is.

Pillbug #1 __________ mm long

Pillbug #2__________ mm long

Is there anything else special about your pillbugs?

Is your race course indoors or outdoors?

If it is outdoors, is it in a sunny area?

Is there anything else special about the race course (texture, moistness, surface)?

NOW...What are your pillbugs' scores?

Pillbug #1________ seconds

Pillbug #2________ seconds

What was the time for the fastest pillbug?

Fastest pillbug ________ seconds

Sarah has watched Patricia in many situations. Which of her observations might be useful when it comes to making Patricia run quickly?

Answer this question *after* the races: If you wanted to get a really fast pillbug, where would you try to find it and what characteristics would you look for?

Fifth Day

Teacher Note

First, have the PILLars, PILLows, and PILgrims go out and check their observation station. If some groups aren't getting any pillbugs, point out that this is, nonetheless, a result—one that can be compared with the results of others. But if there is genuine discouragement on some students' parts, you might encourage them to put out an additional trap or brick or test object. If you started the Pillbug Project on a Monday and have gone straight through, then this is a Friday, so they will be leaving the observation stations out over the weekend. This is excellent, for it will give the pillbugs ample time to find and acclimate to whatever they encounter.

Since today's drawing project should stem from whatever mood the *Patricia Pillbug* story might leave, it would be best if the art supplies are all easily accessible so that the children can immediately get to work without a lot of distractions. You may even want to tell them before starting the story what they are going to do after the story ends. Most of the children will draw something related to the final scene of the story unless you encourage them to think a bit about the whole story before they begin. To remind the students of other parts of the story you could pose such questions as: "How would you feel if, like Patricia, your home was picked up and moved someplace else?" "How would your feel if your home was about to be eaten?"

There is, by the way, no special student notebook page for today. The students' drawings can be put into their notebooks, or could even serve as their notebook covers.

Materials List

- Art supplies—The children will be drawing a picture about the *Patricia Pillbug* story.

Our story continues . . .

We left the story just after Sarah put Patricia in a yogurt container and decided to take her to school where Sarah's class is studying pillbugs. (Can you imagine anything more weird than going to a school where they study pillbugs?) Sarah had decided to call Patricia "Fred." You must understand that while pillbugs can easily tell a boy pillbug from a girl pillbug, it is much harder for people to tell the difference. So it is not Sarah's fault that she called Patricia by a boy's name. Nevertheless, Patricia had every right to be upset, if a pillbug could even be upset. If someone called you yucky, tried to grind you up in a garbage disposal, dropped you in a slimy yogurt container, and called you a boy if you are a girl or called you a girl if you are a boy, you'd have a reason to be upset, too. Surely things were going to get better for Patricia.

The next morning started well enough. Sarah had cereal for breakfast and Patricia had blueberry yogurt. (Sarah's mother said that was Sarah's only choice since they were out of eggs. Patricia had no other choice since she was trapped in only one type of yogurt container.) Sarah wore a blue shirt; Patricia wore a blue stain from the blueberry yogurt.

Off to school they went. Sarah and all the other children listened to a story about a girl and a pillbug the girl found in a strawberry, and then their teacher said that they were going to have pillbug races.

What excitement! The children ran around lifting up sticks, dead leaves, and rocks looking for the best—the fastest pillbug. They tested them; watching them walk in wet places and dry places, in the sun, on a rug, on sand. Sarah just watched the hunt for the fastest pillbug. She remembered how fast Patricia ran from her back at home at the kitchen sink. Sarah decided to let Patricia run last of all. Sarah felt sure her team would win the PILLBUG RACERS' CHAMPIONSHIP CERTIFICATE.

The races began. One pillbug ran around in circles for three minutes. Everyone laughed—except for the team whose pillbug seemed to be getting dizzy. One pillbug appeared to think that the chalk line that was the starting circle was a highway. It climbed on the line and zoomed around it, never getting off. Most of the competing pillbugs finished the race with a score of 80 to 150 seconds. One outstanding pillbug the students called Zappo got a score of 34 seconds. Zappo was the pillbug to beat.

Finally the teacher announced that it was time for the last team, the one Sarah was on, the one Patricia was on. Each team was allowed to race two different pillbugs. Sarah's team first entered a little bitsy Roller that they called Ruth. The team decided that it might run best if it was really happy. So they fed it alfalfa sprouts before the race, kept it in a wet tissue nest, sang songs to it, and in general, were as nice to Ruth as they could be.

They put her in the starting circle and she wandered slowly away. She crossed the starting line. The official timer called out "Start...NOW!" and Ruth curled up, closed her eyes, and went to sleep. Actually, she couldn't close her eyes, but she sure did curl up and stay there. No amount of shouting would make Ruth move. She seemed to be too full of alfalfa sprouts to care about anything but taking a nap. Sarah's team was pretty upset. Patricia was their only hope.

"Are you sure Fred is fast?" asked Sarah's teammate.

"He was fast in our kitchen sink," Sarah assured her.

"He'll never beat Zappo!" shouted someone on Zappo's team.

The teacher announced that it was time. Sarah carefully took off the lid of the yogurt container. Patricia peered up. Sarah said, "Come on Fred, now is your chance."

Sarah tipped Patricia out onto the racecourse, right in the middle of the circle. Patricia was all covered with blueberry juice from the yogurt container. She was slimy with yogurt, she was sticky with yogurt, she had had enough of yogurt.

"COME-ON FFFFREEEEEDDD!"

"GO, BOY, GO!"

"GIDDYAP BUG!"

Almost all of the kids were cheering. Only Zappo's teammates stayed as quiet as rocks in the bottom of a mud puddle.

hat happened next really happened, but different people say different things. Some say it just happened to happen, some say it shows pillbugs are really special. Some say this is just a made-up story... You know if you put a pillbug in water for a few seconds and then let it walk around, sometimes it drags its tail-end on the ground. Scientists say that is a way they dry themselves off when they are really drenched with water. When a pillbug does this, it leaves a wet line behind it. If it walked in a circle, it would leave a large wet "O" on the ground.

Well, Patricia was slimy wet with blueberry juice. She started trudging, wiping off the purple blueberry juice as she trudged.

"YEA! FRED," almost everyone shouted as she crossed the starting line.

"Starting right...NOW!" shouted the official timer.

"NO! GO STRAIGHT," shouted most everyone as she picked up her tail and started off.

"TURN AROUND," shouted Zappo's teammates.

"HE'S MAKING A BLUE LINE ON THE GROUND," shouted one girl who was good at noticing.

Everyone stopped shouting, marveling at the blue trail. Patricia could feel the vibration cease, so she stopped too. After a moment, she set her tail back down and was about to begin hiking again when Sarah shouted "Don't stop, Fred! Run!" Her tail came up, leaving a small blueberry dot. The other teammates joined in, "Run, Fred!" The vibration from all this shouting was not as intense as the garbage disposal had been, but it was enough to start the little hiker pillbug hiking again. She paused twice, picking up her tail for a moment each time and ending the trail of blueberry juice. But the ruckus above her continued, for she was hiking along faster than Zappo. As she crawled across the finish line, three seconds ahead of Zappo's time, Sarah scooped her up.

"Fred, we won, we won! You're the best pillbug ever! I'm sorry I said you were yucky. I'll keep you forever and ever!"

Patricia didn't do anything. She just sat in Sarah's fist, resting. Sarah looked down at the blueberry trail still on the giant sheet of brown paper. It looked like a long dash, followed by a dot, followed by three long dashes.

"Look at that," she exclaimed.

Another classmate, Chris, whooped, "Look, a Morse code pillbug! Do you think it means anything?"

"No," said Sarah's teacher. "Pillbugs don't know words. But you're right, it looks like Morse code. Let's look in the dictionary at the way Morse code represents letters."

Chris found Morse code in the dictionary. "Fred said 'no'," she whispered, looking back at the marks. "Dash dot, pause, dash dash dash means no."

"What?" said Sarah. "But pillbugs don't know words. How could she have spelled the word 'no?'"

"No," whispered Sarah's teacher.

"No," whispered Chris.

"No," whispered the kids in the class. Patricia sat there in Sarah's hand, just being a pillbug.

"What does it mean?" Sarah asked her teacher. People have asked "What does it mean?" before, but never in the history of the world was the question asked about the word "no" written in Morse code and blueberry juice on a giant sheet of brown paper in a classroom.

Sarah's teacher was usually very good at answering students' questions but had to think about this one. "Probably it happened by chance. But maybe we should remember from this that even little pillbugs are alive and maybe in some way care. For a pillbug, a bunch of kids having a race with pillbugs is probably not their preferred way of life," said Sarah's teacher, who then paused and looked off into the distance. Finally Sarah's teacher looked down again at Sarah who was quietly standing there looking down at her hand closed around Patricia. "Where did you say you'd found that pillbug?"

"It was curled up in a strawberry," replied Sarah.

"The strawberry was a kind of home for it, wasn't it? That's probably where it would really want to be—in a strawberry patch—not in a classroom or in a house," said the teacher.

Sarah and her team were given the award for the fastest pillbug.

That night Sarah took Patricia out to the strawberry patch, poked a hole in a bright red strawberry and gently slid Patricia into the hole. "Good night, Fred," whispered Sarah.

That night a tired pillbug, with bits of blueberry yogurt still on it, curled up inside of a wonderfully wet and cool strawberry.

Teacher Narrative

What I would like you to do now is to make a picture of some part of the story about Patricia. Don't draw anything for a few minutes—just sit quietly, and think about some part of the story. Try to get a picture inside your head. Maybe that picture is not like anything you have ever tried to draw before. When you decide what you want your picture to look like, get started.

Teacher Note

If students need more encouragement about what to draw, some questions you might ask are: "During what part of the story do you think Patricia was most in danger?" "How did Sarah find out how fast Patricia could run?"

Sixth Day

Teacher Note

Have your students check their observation stations, record their data, dismantle the stations, and restore any resident pillbugs to a safe area. Then give them time to prepare for their reports. Let them know that they are the experts and that they will therefore be sharing information with the class as authorities. If the tone is set right, many students will respond by giving very competent reports on their findings. If you ask questions based on your interest in their results, and if the students are encouraged to ask similar questions, the talks can be similar in many ways to those given by professional scientists.

Teacher Narrative

This is the last day of your three-day observations as PILLars, PILLows, and PILgrims. Go out and check your observation stations, write down today's results, and then dismantle the set-ups and restore any resident pillbugs to a safe area. When you come back we will prepare for our first symposium.

Teacher Narrative

Now it is time to have our first symposium on your Pillbug Project research. First, I am going to give you the guidelines scientists use to give reports to their colleagues, and then you will have about 20 minutes to get ready. Then three people from the PILLars, three people from the PILLows, and three people from the PILgrims will present their research to the rest of the class.

How scientists give a report

People conduct science experiments in teams all over the country, all over the world. Scientists work for companies; for museums; for city, county, state, and national governments; for colleges and universities; for special research institutes; and for hospitals. Scientists work in a lot of places, discovering new things and new ways to use that new knowledge. But no one scientist, or even one team of scientists, can know everything about a subject.

So scientists spend a lot of time communicating with each other—keeping up with the latest discoveries in whatever area they are working in. Scientists studying how clouds form talk to and read reports by other scientists studying weather systems. Scientists studying how to make metal stronger and lighter for cars talk to and read reports by other researchers and technicians who study metals.

We have an image of the scientist of a hundred years ago as a lonely hermit, working away in a tower, perhaps trying to turn lead into gold or make people fly. Today's scientists have much more fun: They work together to find where gold has been created by the Earth's processes. They create and assemble technology to allow people to fly with bicycle-powered airplanes, with jet-powered backpacks, with planes powered by propellers on top, in front, or behind. And they discover how tiny creatures eat and move and protect themselves and find safe places to live.

Now we, as famous pillbug researchers, are going to share what we learned in our Pillbug Project. First, let's talk about what scientists need to do so they can communicate quickly and efficiently. Remember, a scientist has to divide his or her time between communicating—sharing knowledge with others—and doing research.

1. Scientists arrange their work and results so others can logically follow it. They are organized.
2. Scientists start by explaining what their research question was—what they were trying to find out. If they were working on spiders in a house they might say, "We wanted to see whether spiders live mostly in rooms where people have food."
3. Scientists then say briefly what they did to answer their question. The spider scientists might say, "We counted all the spiders we saw on the walls and the ceiling of each room of the house. We measured each room so we could say how many spiders there were for each cubic meter of space in each room." (They don't have to say how they walked from room to room or stuff like that.)
4. Then scientists give the numbers and facts that they found. Often it is a good idea to write on the blackboard the numbers and where they came from because most people can't remember a bunch of numbers. The spider scientists might write down:
   Kitchen (36 cubic meters)—27 spiders
   Dining room (38 cubic meters)—34 spiders
   Bedroom (42 cubic meters)—9 spiders
   Bathroom (12 cubic meters)—2 spiders
5. Then the scientists say what they think their numbers and facts tell them about the animal they are studying. Other people may think their facts show something more or something else. Our spider scientists might say, "We think the reason we found more spiders in the dining room and the kitchen than in the bedroom and the bathroom is that spiders go where flies are, and flies go to the dining room and the kitchen where food is." If there are other possible reasons for what they found out, they might point them out or tell why those reasons aren't good ones.
6. Scientists end by repeating what they really want people to remember about their work. They might say, "So you see, we have checked and found that there are more spiders in rooms where there often is food. We think that is because those rooms have odors that attract flies, and spiders go where there are the most flies."

So let me now repeat what you should remember about giving your reports on your observations:

- Start by describing what you were trying to find out.
- Tell the important part about what you did to answer your question about pillbugs.
- Give the facts—in this case, how many pillbugs you found in each place you tested.
- Explain what you think your observations tell us about pillbugs. Summarize.

The PILLars, the PILLows, and the PILgrims should get together in their groups. Get all your results organized, review what you have recorded in your research notebooks, and talk about your results. Try to decide what you found out and what it means. You may decide to talk about only some of your results. That is fine.

Your group should pick three people to talk:

- One person will tell what the observation station was supposed to find out and how you made your observations.
- The next person will explain what your group found out—just the facts. Write your numbers on the blackboard and write a few words next to each number (such as: 47—number of pillbugs under a small board).
- The last person will say what your group thinks your work shows and then he or she should repeat whatever it is that the group thought was the most important part of their work.

Please remember—you scientists are the experts and this is your symposium. No one in the world has done your exact investigation or knows what you found. When you are talking, speak slowly and clearly.

When you are listening, listen quietly and carefully—you will learn. Ask questions of the people who give the talks—that is how you can learn more. You are going to have a meeting of scientists—yourselves—just as scientists 30 years older than you do all over the world.

Symposium #1

Look at the data table you and your group members have made and the questions you answered. Talk about them. Decide what your group found out and what each discovery means. You may decide to talk about only some of your results. That is fine. Use the summary below as a reference while preparing your report.

Parts of a Research Report

- Organization—Arrange your work and results so others can logically follow it.

- Purpose—Start by explaining what your question was—what you were trying to find out. If a group of scientists were working on spiders in a house they might say, "We wanted to see whether spiders live mostly in rooms where people have food."

- Procedure—Explain briefly your procedure, what you did to answer your question. The spider scientists might say, "We counted all the spiders we saw on the walls and the ceiling of each room of the house. We measured each room so we could say how many spiders there were for each cubic meter of space in each room."

- Observations—Give the numbers and facts that you found. Often it is a good idea to write on the blackboard the numbers and where they came from because most people can't remember a bunch of numbers. The spider scientists might write down:

  Kitchen (36 square meters)—27 spiders
  Dining room (38 square meters)—34 spiders
  Bedroom (42 square meters)—9 spiders
  Bathroom (12 square meters)—2 spiders

- Conclusion—Say what you think your numbers and facts tell you about the pillbugs you are studying. Other people may think your facts show something more or something else. Our spider scientists might say, "We think the reason we found more spiders in the dining room and the kitchen than in the bedroom and the bathroom is that spiders go where flies are, and flies go to the dining room and the kitchen where food is." If there are other possible reasons for what you found out, you might point them out or tell why those reasons aren't good ones.

- Summary—Wrap up your presentation by repeating what you really want people to remember about your work. The spider scientists might say, "So you see, we have checked and found that there are more spiders in rooms where there often is food. We think that is because those rooms have odors that attract flies, and spiders go where there are the most flies."

Your group should pick three people to talk:

- One person should tell what your group was trying to find out. Then that person should tell what you did to answer your question about pillbugs.

- The second person should explain what you found out—just the facts. Use the blackboard to write how many pillbugs you found in each place you tested and a few words next to each number (such as: 47—number under a small board).

- The third person should say what your group thinks your investigation shows and then should repeat whatever it is that you think is the most important part of your investigation.

You scientists are the experts. No one in the world has done your exact investigation or knows what you found. When you are talking, speak slowly and clearly. When you are listening, listen quietly and well—you will learn. Ask questions of the people who give the talks—that is how you can learn more.

Symposium Report Notes

What we were trying to find out:

What we did to answer our question:

What our observations were:

What our observations tell us about pillbugs:

Summary:

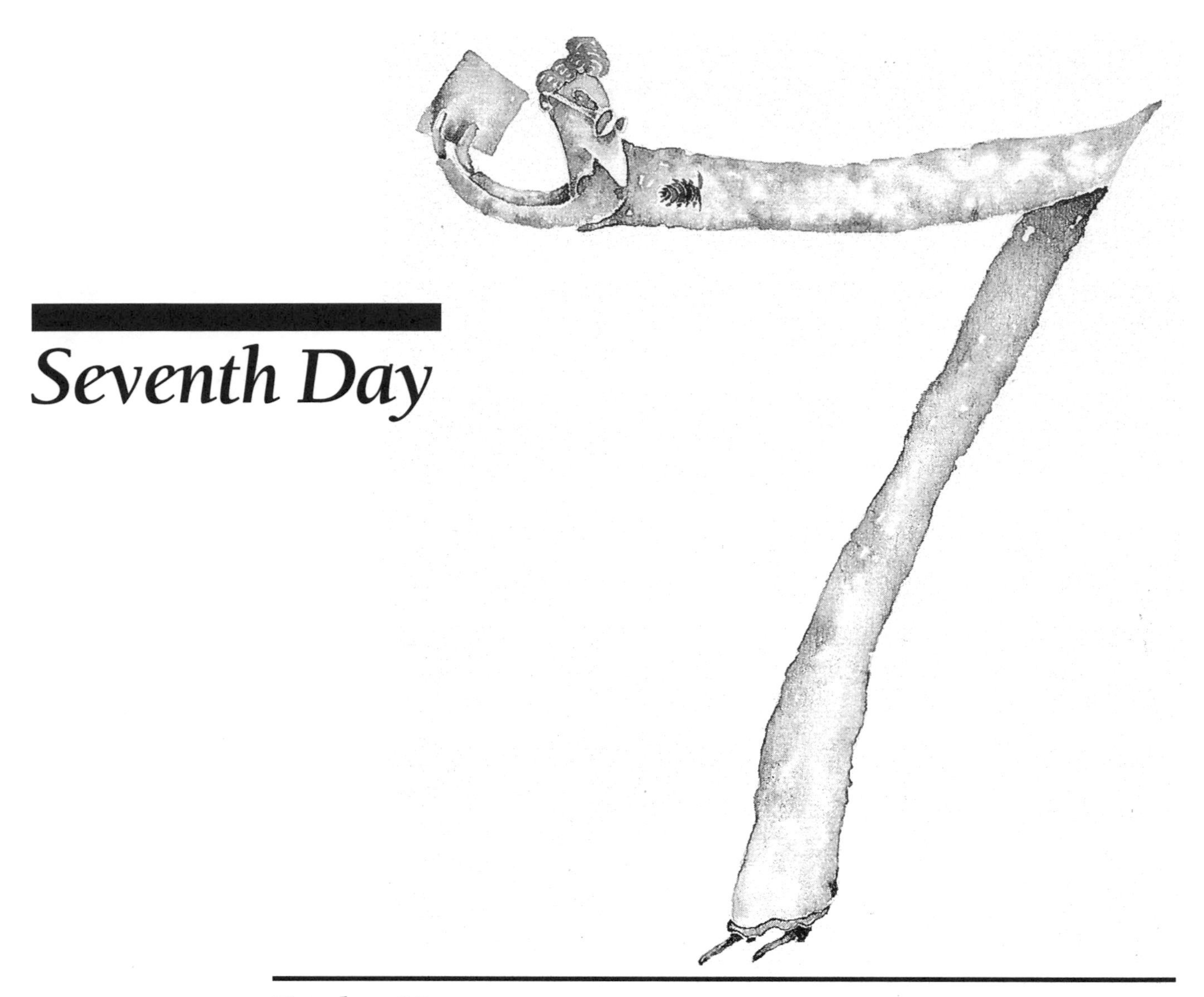

Seventh Day

Teacher Note

Today you are going to start by encouraging the children to see pillbugs from the pillbug's perspective. What could be an obstacle for a pillbug? What could be dangerous? The scene below takes your listeners on a short journey to find out and suggests how a pillbug might respond to these situations. Today's activity invites the students to further think about a pillbug's world and how a pillbug's structure and behavior—which the students have been examining these past few days—might help them survive.

Take three or four large sheets of newsprint paper (however many groups you want to divide your class into) and write a question at the top. Mount them at stations around the room. You will need the same number of stations as groups of students. It will be easiest to prepare these sheets and have them set up ahead of time. The following questions can be used, or you may want to come up with your own that require students to think about the function of a pillbug's structure or behavior in its survival.

- How would multiple legs help you survive? How would they be harmful to your survival?
- How would overlapping scales help you survive? How would they be harmful to your survival?
- How would antennae help you survive? How would they be harmful to your survival?
- How would rolling up like a ball help you survive? How would it be harmful to your survival?
- How might a pillbug's life be different without birds?
- How might a bird's life be different without pillbugs?
- How might your life be different without pillbugs?

At each sheet there will be one group of students. Give each group a different colored marker, so you will know who wrote what answer. Allow 5–10 minutes for the groups to answer their respective questions, either by writing, drawing pictures, or both. Then have each group move to another sheet. When groups arrive at a new sheet, they must read not only the new question but also all the previous groups' answers to that question, because their answer needs to be different. Continue this procedure until all the groups have answered all the questions.

This is a method of project assessment called a gallery walk. This activity should serve several functions—to assess what students are learning from the Pillbug Project, to build student enthusiasm as they see each other's responses, and to make students consider that there are several answers to questions in science.

Materials List

- Large sheets of newsprint paper—Each group of students will need space to draw on here. You need as many sheets as there are groups of students.
- Colored markers—It works best to have a different color for each group so you know which group wrote which answer.

Teacher Narrative

Imagine being a pillbug. Really. Close your eyes, and try very hard to think of yourself as a pillbug. Think of being very small. Think of having no hands. Think of having legs all over. Sit quietly for a little while and really imagine being a pillbug…

Look at you—you are really cute as a pillbug with your little bitty black eyes and quivering antennae. Imagine crawling down a gravel path towards some pretty flowers. Each piece of gravel is now a huge boulder. You are going to have to climb up and over that one in front of you. You don't have hands so you can't grab on—just stick your little legs in any teeny crack you can find. What looks like a smooth surface to a person is to you almost like a ladder with little notches and cracks for your legs. Start climbing.

Now, high in the sky a black cloud drifts over you. A large raindrop starts falling, faster and faster, falling straight toward you. You don't see it coming. Just as you reach the top of the piece of gravel, the raindrop hits you. Imagine this glob of water as big as you are smashing into you after falling for kilometers. Imagine the sound, the surprise, the wetness! It knocks you off the piece of gravel, and you fall to the ground below you, landing on your back.

Hurry! Wiggle your legs, arch your back, try to turn over! The shadow of a bird crosses over you as you struggle. You look up and dimly see the bird flying towards you. He lands right next to you and starts to tilt his head to look you over. Stop moving. Birds sometimes don't eat dead pillbugs, and you will look a lot more dead if you hold still. The bird opens his beak. He is getting ready to jab you and pick you up and swallow you.

But wait! A black dog sees the bird and starts chasing it. You are saved. The bird takes off, and the wind from his wings tumbles you over and over. Each time you roll, your head crashes into the ground. The dog charges by you and one of its paws just misses you and pounds into the ground next to you sending bits of sand and dirt flying in all directions. From your perspective, each grain of sand is a rock able to break a leg.

But the flying sand misses you, the dog is gone, the bird is gone, and you can get back to walking.

Now it is time for *you* to do some walking. You have probably noticed the large sheets of paper stationed around the room. We will divide into groups and each group will have its own home base at one of these stations. Each sheet has its own question on it. As a group, come up with an answer to that question and write or draw your answer on the paper. But leave room on it for all the other groups to respond to the question as well. You will have XX minutes to put your answer down, then you will move to a different question. (Teacher: Adjust the time allowed for student age and group size.)

When you get to a new question, read the answers the groups before you wrote down. Yours must be a different answer. All the groups will have an opportunity to answer all the questions.

Teacher Note

A bit more on the background of the gallery walk as a form of assessment. The goal of this assessment is to determine how effective a student is in applying his or her ideas when solving a problem. The group interaction of this day's activity both challenges ideas as students raise them and reinforces the validity of ideas as students watch them pass the test of their peers. The social dialogue also forces students to think more deeply about their ideas, resulting in better ideas. In this way, the cooperative assessment used in Seventh Day is as much a learning process as a product.

Eighth Day

Teacher Note

Today's beginning discussion addresses the question "Are you testing what you think you're testing?" Technically, this is the concept of validity, though you don't need to use that word when you are presenting the concept to your class. Students are often not aware of how many variables can affect the results of a scientific investigation. Consequently they don't test what they intended to. The narrative in today's project sets up comic situations for experimental design, with questions to inspire class discussion on how to design a valid experiment. Use what you think will best help students grasp this concept. You will want to reinforce this concept of validity each time you teach science projects.

Materials List

- A plastic container for each team—There are many things you can use as the containers for the investigations on choice. Deep plastic margarine tubs or the cut-off bottoms of plastic gallon milk containers are two inexpensive solutions. Or you could buy cheap plastic containers from a restaurant supply or janitorial supply store. Plastic is good because pillbugs can't climb up the sides.

- Paper towels or plaster of paris—These are two options for what to put in the bottom of the test container. The quicker, cleaner approach is to tape a paper towel to the bottom of the container. The tape keeps the pillbugs from crawling under the towel, avoiding the test altogether. Be careful that no sticky tape is exposed; it will act like fly paper and pillbugs will get stuck.
- Plastic wrap—To cover the test containers to keep the pillbugs moist inside.

Teacher Note

If you have the time and energy, the plaster of paris provides a better test chamber. Plaster of paris makes a uniformly damp bottom that holds test liquids well, but it takes more time to set up. If you use plaster of paris, make sure you have enough on hand so that each container has one to two centimeters of plaster on the bottom. Mix it so that it is fairly stiff—almost as thick as biscuit dough. If it's too thin it will shrink away from the sides when it dries, and the pillbugs will crawl underneath and not choose anything, except to be out of the way. Make sure that after the plaster is added to the containers, the containers are tapped on a table top several times so the plaster settles into a flat surface.

There should be at least 15 pillbugs for each team of two students. If pillbugs are scarce, you may want to group students into larger teams.

Make sure that each team of students selects something reasonable to test. "Reasonable" can be given a fairly wide latitude, however. One student put Tabasco sauce on one side and garlic powder on the other. While it may not be clear what the results implied for a pillbug in the real world, at least the pillbugs made a clear choice between the two. The more interesting observations tend to be ones that say something about a pillbug's real world—like choosing between wet sawdust and dry sawdust. But the students are the scientists now. Encourage them to measure how much of each thing they test. There may be a difference to the pillbugs between diluted vinegar and pure vinegar, or between moist oatmeal and very wet oatmeal, so the students should know how much water or vinegar they used so the observation could be replicated.

You should make sure that whatever is to be tested tomorrow is indeed there tomorrow. Much of it is easy for you to bring from home, and much can probably be found around school. But if a student selects something special, you and the student must make definite plans to get that something to school tomorrow.

Teacher Narrative

Later today you are going to give pillbugs a chance to do some choosing. You will let them choose between being in the dark or being in the light. Or choose between cream of wheat or sawdust, for example. This type of test is called a "choice experiment."

First, you should know why and how scientists do choice experiments. Next, you should know some things scientists check for when they design an experiment to make sure they are testing for the right thing.

First, why do scientists do choice experiments? Scientists have a lot of reasons for doing choice experiments. One scientist may be interested in making a new type of trap for flies, for instance, and wants to find out what is the best bait for the trap. The scientist could put flies in a container that has fish on one side and rotten cottage cheese on the other.

If the flies all choose to go to the fish, what could the scientist conclude?

What if the fish was warm and the cottage cheese was cold?

Could you be sure whether the flies were attracted to the food or to the temperature?

How could you rule out temperature?

Now suppose you, or any other scientist, has a question about what an animal chooses, and so you decide to do a choice experiment. You must be careful that your test is testing what you really want it to test (that it is *valid*).

Let's consider an example. Pretend the cooks in our cafeteria want to know whether they should serve pizza hot or cold, so they enlist the help of some cafeteria consultants—Beatrice Ware, Louis Nee, and Moses Ronn to do an experiment. Now, these consultants, better known as Bea Ware, Lou Nee, and Moe Ronn, decide to give a group of students a choice between a hot food and a cold food, thinking that temperature is the only important thing to vary and the type of food they use really won't matter. So they let the kids choose between hot liver and cold pizza. All the kids except for one, who likes being different, pick the pizza. Bea, Lou, and Moe decide from their test that children like cold food more than hot food, so they told the cooks to serve cold pizza instead of hot pizza.

These cafeteria consultants thought they were testing a choice between hot and cold, but what they really tested was a choice between liver and pizza. It's a safe bet the students would have chosen hot pizza instead of hot liver.

You guys know that Bea Ware, Lou Nee, and Moe Ronn made a mistake, even though their test was done carefully. Their test didn't test what they thought they were testing. If the cafeteria hired them to see if children liked pizza hot or cold, what test should they have done?

Let's look at another example. This time Lou's brother Zane and his friend Dewey Fuss are hired by a company wanting to make earwig traps. (Earwigs are 1–2 cm long insects often found in moist woods or gardens.) They want to know if earwigs prefer dark or light places. So Zane Nee and Dew Fuss put a light bulb in one end of a long box and add a bunch of earwigs. All of the insects move away from the light to the other end of the box. Based on this observation, Zane and Dew tell the company that earwigs choose dark over light.

Since we don't know as much about what earwigs like as we do about what people like, it is harder for us to see what could be wrong with the test.

How could they be wrong?

What else in the box could the earwigs be moving away from?

Maybe they were moving away from music that was coming from a radio near the box. Or maybe they moved away from the light bulb because it was hot, not because it made light. In other words, maybe Zane and Dew were—like Bea, Lou, and Moe—testing more than one thing at a time and the earwigs responded to something other than what Zane and Dew thought they were responding to.

OK scientists, the rest of today's part of the Pillbug Project and all of tomorrow's part of the Project are up to you. You folks are the scientists. Today you and your team are going to build special test containers for your pillbug subjects and think up tests to find out what pillbugs like or don't like. Your Research Note page for today has instructions for building your container and a list of possible tests you may choose. If your team would like to test something not on the list, check with me to make sure we can get the materials for the test.

Tomorrow you will begin with quick tests to see which items you decide to test might be most valuable in learning about pillbugs. Then you will carefully set up that one test with a choice of two different things for the pillbugs to walk on or go under, and add 15 pillbugs. You'll leave them alone for 30 minutes, and then you will count how many pillbugs are on or in each of the two things. Then you will leave them alone overnight and check them again the next day.

When you're conducting your observations, you need to keep an important point in mind: What is harmful to you may be harmful for pillbugs. Remember that you are testing their preferences only and you want them to be safe and healthy.

What Will Pillbugs Choose?

What to do

1 You will work in teams. Each team will get a container. Your teacher will tell you how to prepare your team's test container. You will use either a paper towel or plaster of paris on the bottom of your container to keep the pillbugs moist during your test. Put the names of the team members on the container.

Paper towel procedure—Carefully tape a paper towel along its entire edge inside the container at the bottom. The tape keeps the pillbugs from crawling underneath the paper towel, avoiding your test altogether. Be careful that no sticky tape is exposed; it will act like fly paper and pillbugs would stick to it. Put a few drops of water on the paper towel and cover the container with plastic wrap. You might need to reapply water to the towel each day to keep it moist, but not wet.

Plaster of paris procedure—Carefully put enough plaster of paris in the bottom of your container so that the layer is 1–2 cm deep. Plaster holds water for a long time, so it will keep the pillbugs from drying out during the test. After you have poured the plaster into the container, tap the bottom of the container gently on the table several times to knock out air bubbles and clean off any wet plaster that might have hit the sides (pillbugs could climb up the plaster and get out). Tomorrow, when the plaster is set, paint water on the plaster and cover the container with plastic wrap.

2 Make sure you keep the pillbugs moist during your test so they stay healthy.

3 Choose two things that a pillbug might normally have to choose between. Below is a list of some of the things you could choose. If you think of something else you would like to test, check with your teacher to make sure it is OK and if it is, give it a try.

Here is a list of things pillbugs might choose from.

- wet corn meal on one side, and wet sawdust on the other
- cut-up strips of cardboard on one side, and cut-up strips of plastic bag on the other

- a piece of clear plastic bag they can climb under on one side, and a piece of dark plastic bag they can climb under on the other
- using a paintbrush, paint a thin layer of very salty water on one half, and paint plain water on the other
- paint a thin layer of vinegar on one side, and water on the other
- paint a thin layer of fruit juice or hot sauce or juice from another kind of food on one side, and water on the other
- dirt on one side, and oak leaves on the other
- paint one side one color, and the other side a different color
- coffee grounds on one side, and cereal on the other
- a place with holes that a pillbug can crawl into, and a place that has no holes
- anything on one side, and nothing on the other
- Team suggestion:
- Team suggestion:

Bring from home anything you want to test that isn't already in the classroom. Check with your teacher first to see what is already available.

When you're designing your tests, try to remember about testing just for the thing you want to test for. If you are wondering whether pillbugs like sweet things more than salty things, don't put watermelon on one side and a pile of salt on the other—they may pick wet things over dry things, or solid things over loose things, or red things over white things. Set your test up so that there isn't more light or sound or wind or heat or moisture on one side than another, unless one of those things is what you are testing.

4 Fill out this data before you do your test.

What items will my pillbugs choose between?

Which do I think my pillbugs will choose? Why?

Can you relate the choice made in your container to a choice a pillbug might make in its natural habitat?

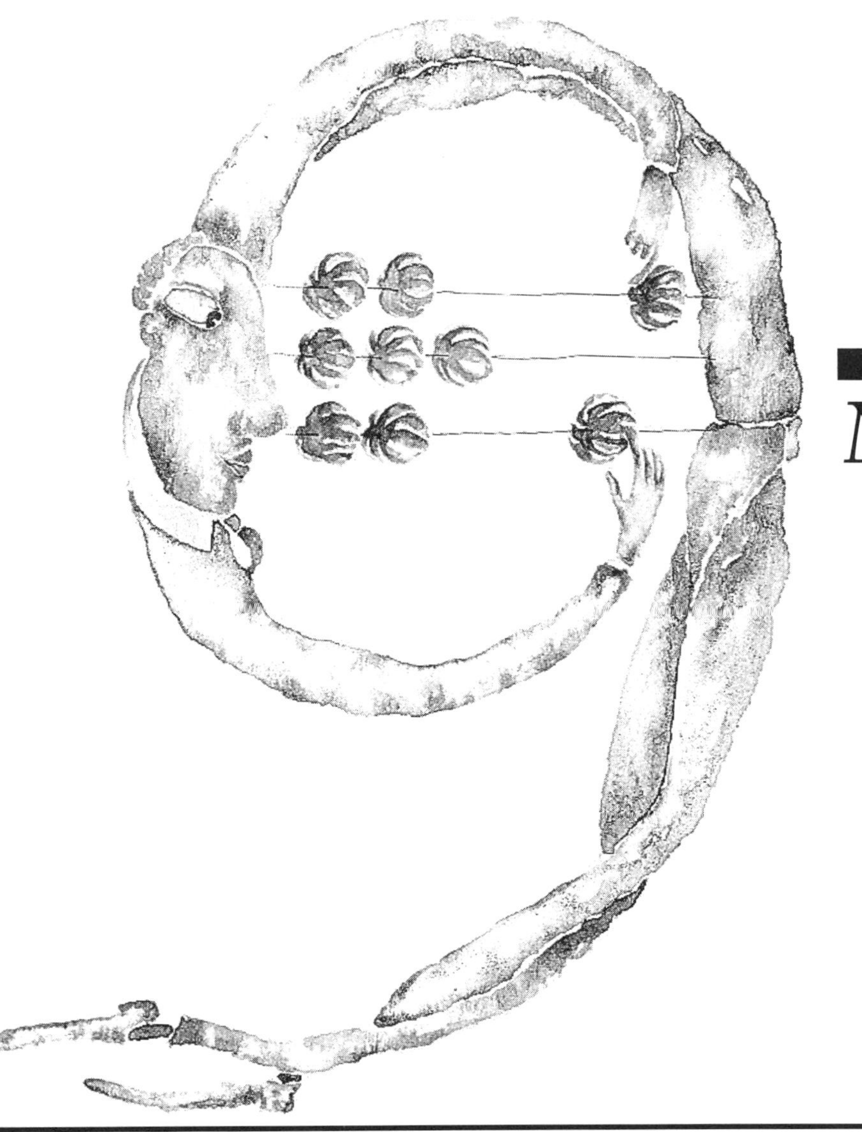

Ninth Day

Teacher Note

Before beginning the day's investigations, you will present another concept—statistical significance. (You won't have to use the words "statistical significance.") Many students don't realize that if eight pillbugs choose red and seven pillbugs choose green, it doesn't prove that pillbugs prefer red over green. Each student will take something different away from this session, and that is fine. Many won't be ready for these concepts, but by presenting them, you are preparing them for a later time when they can grasp these ideas more completely. The narrative in today's project again sets up a comic situation for experimental design, with questions to inspire class discussion of solutions. Use what you think will best help students grasp this concept. You will want to reinforce this concept each time you teach science projects.

This is the day for the choice experiments. After the brief discussion, you'll just need to encourage your students to get started, and then move around supplying them with materials and helping them cope with any dilemmas they encounter. It is a good idea to have some reasonable items available to use in the choice experiments for those teams who forgot to bring anything or who find that their original experiment isn't going well. I don't want to give results away, but perhaps you should know that pillbugs seem to be selective about sour things, amount of light, and moisture (and other things yet to be discovered).

Materials List

- Test containers—The ones the students made for Eighth Day.
- Test items—Whatever students selected on Eighth Day to test today.
- Pillbugs—At least 15 for each team of scientists. Each team should use either all Rollers or all Hikers, not a mixture.
- A spoon for handling pillbugs
- A pencil
- A paintbrush—For applying test substances to the bottom of the test container.

Teacher Narrative

Before you get started on your observations, we'll talk briefly about how to tell how sure you are about what you found out (your results).

Imagine the telephone rings, the principal answers it and finds that it's the president of the United States. He has heard that you people are experts on pillbugs, and he has a problem he needs you to solve. Pillbugs keep crawling into the space shuttle just before it is supposed to take off. They crawl into the computer that controls the space shuttle and keep it from working. The president thinks they do this because the computer is colored green like the grass they walk around in. Other people think they are climbing in the computer for other reasons. He has to find out quickly whether they are attracted to green things so the computer can be painted another color, if necessary, and the shuttle can take off. So the president asks you to test your pillbugs to see whether they prefer green to another color.

Well, you find 20 pillbugs and put them in a box that has green paper on one side and red paper on the other. You leave them alone for an hour, and then you look in the box. What if all 20 were on or under the green paper? You'd probably call the president and tell him to paint the computer red and get ready for lift-off.

And if there were 10 pillbugs on the red and 10 on the green? You'd probably call him and tell him that pillbugs don't seem to care what color they are near.

But what if 12 were on the green and 8 were on the red? Is it clear that pillbugs choose green more than red if you only have a few more on the green side than on the red? Do you think it's possible that an hour later there may be more on the red side than on the green?

What would you think if there were 16 on the green and only 4 on the red?

How many would have to be on the green side before you were pretty sure that pillbugs chose green instead of just happening to have a few more there, and maybe later there would be more on the red side? That is not an easy question to answer.

Scientists have rules that they follow when they try to decide if what they have found is significant. Scientists use tables of numbers and do special calculations before they can say how sure they are about what their observations show. These calculations are called statistics. If you found that 15 of your pillbugs went to the green and 5 went to the red, and if the results were the same after checking several times, statistics would tell you that you could be pretty sure the pillbugs preferred green. But if only 14 went to the green, the rules say that you can't be sure enough to say that they do prefer the green to the red.

We won't go into any more detail about statistics. Just remember to be careful about saying that the pillbugs liked one thing more than another. If many go to one thing and none to the other, you can be pretty sure, but if only a few more go to one thing than to another, you can't be sure at all.

So, in any investigation you do, remember the two things that we have talked about these past two days.

1. Make sure your test is testing what you want it to. Don't do as Bea Ware, Lou Nee, and Moe Ronn did and test children with cold pizza and hot liver, when you really want to know whether they like pizza hot or cold.
2. Be aware of how sure you are about your results. Don't make the President of the United States paint the computer just because 11 pillbugs chose green and only 9 chose red.

Teacher Note

For older students you may want to add the concept that several trials are necessary to be sure of a fair test. You could also add the idea of more subtle variables such as age of the pillbug, time of day, how soon after eating, etc. will affect results.

What Do Pillbugs Prefer?

What to do

1 With a pencil, draw a line across the middle of the paper towel or the plaster of paris. Write a big "A" on one side of the line and a "B" on the other side.

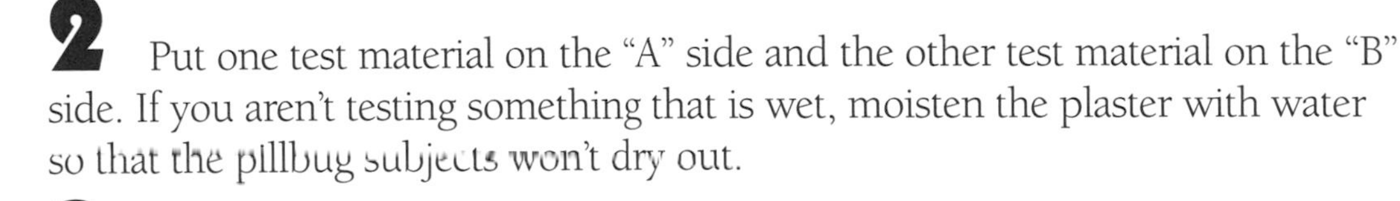

2 Put one test material on the "A" side and the other test material on the "B" side. If you aren't testing something that is wet, moisten the plaster with water so that the pillbug subjects won't dry out.

3 Spoon 15 or so pillbugs right along the middle line. Use Hikers or Rollers, but not both in the same test.

4 Put the container in a quiet spot for a while with a piece of plastic wrap on top. Make sure one side doesn't get more light or heat or noise than the other (unless you are testing one of those).

5 After a few minutes, check your container. Count how many are on each side and write your results in your data table.

6 Check your container after about half an hour. Count how many are on each side and write your results in your data table. If most of the pillbugs are on one side and you are happy with your container, leave it alone until the next morning, and then count them once more. You may want to change your test slightly if the pillbugs aren't making choices or they look like they aren't safe and healthy. (Remember to keep them moist.) If you want to change your test, do so and then leave it overnight.

It is OK if the pillbugs don't do what you expected. Remember that no matter what the pillbugs do, they are telling you something.

What are you trying to find out?

What two things did you put in your test container?

On side A was

On side B was

How many pillbugs did you put in your container?

Were they Rollers or Hikers?

When did you first check your container? After _______minutes

How many pillbugs were on side A?

How many pillbugs were on side B?

Do you think that since there were more pillbugs on one side than the other that you can say that pillbugs will usually choose that side?

What did you find out from this exploration?

What could you do to be more sure of your results?

Write down anything else you found out in the space below.

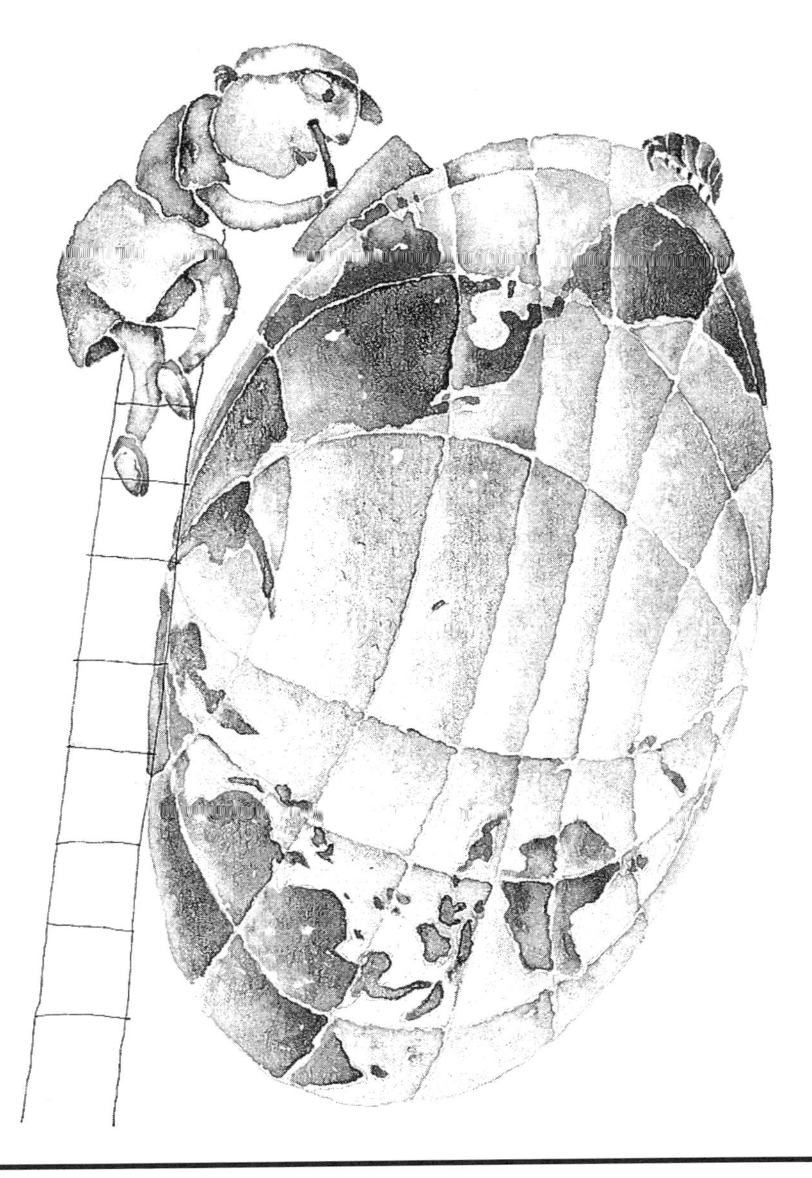

Tenth Day

Teacher Note

This is the last day and the easiest day of all for you. Basically, you encourage the students to organize their results along the guidelines given on Sixth Day. Make it clear that in order for each team to have time to present their results, each report must be brief and to the point. You might put up a large sheet of kraft paper and in big letters make a table containing the names of the team members, the items tested, and a space beside the items for each team to write down the number of pillbugs that chose that item. If you arrange the order of the entries so that related topics are near one another, it is easier to see consistent trends and interesting exceptions. For example, you might group all of the food items together and all of the non-edible items together and so on. The students will be proud of their results, and if you leave the paper up for a couple of days, you might find them discussing their findings further.

Today's Research Notes page has suggestions for further exploration. If students are interested and the pillbugs are healthy, you may want to consider an ongoing pillbug exploration corner in your classroom. These suggestions could also inspire a science project or two.

Teacher Narrative

This is the last day of the Pillbug Project. Let's start by inviting a bunch of pillbug experts for a second symposium about what pillbugs choose. Who are the experts? you ask. They are already here—all of you. It may sound like a joke, but it really isn't. Everyone here now knows far more about pillbugs than do most of the people in the world. Out of every 10,000 people in the world, you know more about pillbugs than 9,999 of them. And during your exploration you may have noticed some things about pillbugs that no one else in the world has noticed.

So, each team will tell the rest of the class what they tested, how long they ran the test, how many pillbugs chose one thing and how many chose the other, and what they think that shows. You'll follow the same guidelines we discussed on the Sixth Day on preparing a research report. Try to keep your talks quite short and only say what you feel is important.

Teacher Narrative

Well, this brings us to the end of the Pillbug Project. Did you have fun? Did you learn something about pillbugs, about science, and about how to learn by looking and by asking and answering your own questions?

This may be the end of the Pillbug Project, but surely you aren't done looking at pillbugs and learning about them and from them. Science investigations often raise more questions than they answer. There are scads of other questions about pillbugs. How could you get the answers to those questions? For some, you could just watch the pillbugs (to answer the question, "Can pillbugs turn over?"). For others, you would go outside and check (to answer the question, "Are there more pillbugs under pine needles than under rocks?"). For still others, you would have to set up observation stations outside—to answer "Are there more pillbugs walking around in a lawn or in a weedy field?" (Remember you used glasses as pitfall traps.) And for some questions, you might decide to take the pillbugs inside and test them (as you did for your choice experiments). You will get one final handout. It has just a few more research ideas in case you want to explore a little more. Add ideas of your own.

What other living things would you like to investigate?

Teacher Note

And thank you. I really would appreciate hearing from you about how The Pillbug Project fared in your class. I think it is a good project, but it certainly can be made better. If you tell me of problems or indicate ways that you improved the project, then the next edition will benefit from your experience and wisdom.

Sincerely,

Robin Burnett

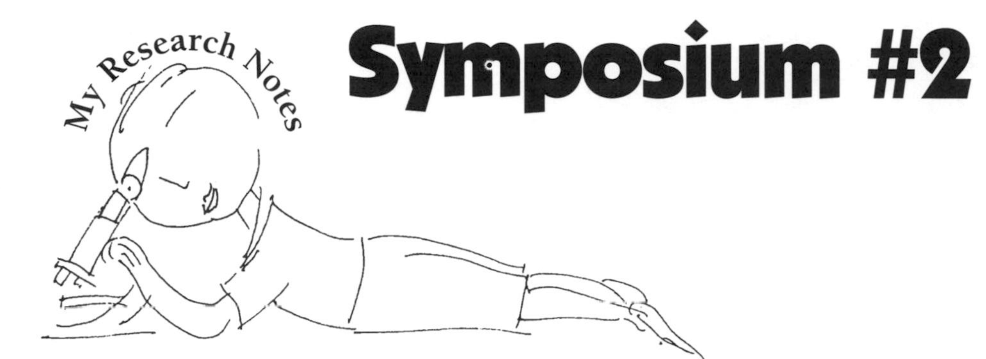

Symposium #2

What to do

Refer to your notes on Day Seven for guidelines on preparing your research report.

Symposium Report Notes

What we were trying to find out:

What we did to answer our question:

Our observations:

What our observations tell us about pillbugs:

Summary:

Future Research Ideas

Here are some questions you might want to try to answer on your own.

1 How many times during an hour does a pillbug crawl out from under wherever it is hiding?

2 Does a pillbug have a home that it returns to, or does it go to a new place every day? You could mark all the pillbugs under a board with a dot from a waterproof marking pen and then see whether you find the same marked ones day after day.

3 When would a Hiker have an advantage over a Roller? When would a Roller have an advantage over a Hiker?

4 If you go on a trip, look for pillbugs. Are Hikers or Rollers more common? Are the pillbugs different in any way from those you find near home?

5 Do pillbugs stay away from places where certain other animals are? For example, do they leave their hiding place if ants or a lizard or a stink bug crawls into the same place?

6 Can you raise baby pillbugs? Try putting a pillbug (one that has a cream-colored patch of eggs in a flat pouch on her belly) in a container with moist plaster of paris on the bottom. Feed her things such as cooked lettuce and cream of wheat. The babies will be very small when they first come out, so plan to look very carefully for them, maybe with a magnifying glass.

7 Do pillbugs like to get together in groups? You could answer this by observing inside or by observing carefully outside. Be careful. There may be several pillbugs next to each other because they are all crowding around the same thing, not because they want to be next to each other. See if you can find out whether they will huddle up next to each other when there is no other reason for them all to be at the same spot.

8 Pillbugs pack whatever they don't digest into flat black specks that look kind of like pieces of pepper. If you feed them food mixed with red food coloring, is that speck red? Do they make more specks when they are dry than when they are kept moist? Do they make more specks when fed one kind of a food than another kind? Can you think of other interesting things about these specks to test?

Try to think of some of your own questions that could be answered by looking or observing. Keep on looking at and playing with pillbugs. They are special animals. Don't forget that there are many other kinds of interesting animals. This project could have been the Cricket Project or the Earwig Project or the Snail Project or the Ant Project—almost any animal would do. Pillbugs are very special, but maybe they are no more special than any other animal. Maybe they seem a bit special because you took the time to be with them for a while.

Pillbug Racer's Championship Certificate

This certificate is hereby awarded to honor

____

as the Fastest Pillbug.

____

____

____

____

(Team member signatures)

____

(Teacher signature)